D1111851

Test of FAITH

Test of FAITH

Spiritual Journeys with Scientists

Edited by
RUTH BANCEWICZ

WIPF & STOCK · Eugene, Oregon

TEST OF FAITH
Spiritual Journeys with Scientists

Wipf & Stock
An imprint of Wipf and Stock Publishers
199 W. 8th Ave., Suite 3
Eugene, OR 97401

www.wipfandstock.com

ISBN: 978-1-60899-894-4

Cover design by Contrapositive
Print Management by Adare
Typeset by Waverley Typesetters, Fakenham, Norfolk

Contents

Foreword

Had I not been so gormless[1] ("cack-handed," to use my parents' picturesque word) with regard to practical things – specifically some experiments called Elementary Analysis – I might have been a scientist myself. Certainly, my chemistry teacher leaned on me very hard indeed to get me to choose science rather than Greek for my ongoing high school education. But I suffer from some sort of disconnect between eye, brain and hand: with wearisome regularity I fail to see what others see needs to be done, and I never seem to do it properly when it is explained to me. As C. S. Lewis could never manage to drive a car, so I am a lost cause on all lab and bench skills, as was already clear to me at age thirteen when the fateful choice had to be made. So it was Greek for me by default, and about the nitty-gritty of the sciences I today know virtually nothing.

Yet I read the chapters of this book with very great interest. These scientists, like others I have known well, seem to me to excel in love of truth, honest humility, and down-to-earthness as a habit of mind. And one of them voices a belief that I have held in quietness for over sixty years and thought was peculiar to myself – namely, that when in the name of science people attack Christianity in savage and sarcastic terms, it is not because they have overwhelming arguments to deploy, but because they have in some way been hurt by persons who professed a Christian identity, and in consequence they are now gripped deep down, deeper perhaps than they themselves discern, by the passion that the world knows as revenge. Be that as it may, across the board, as it seems to me, the level of human insight matching scientific expertise of these chapters is high, and they have left me feeling that here are folk with whom, as a mere theologian, I would love to spend some time. I hope other readers will enjoy this set of testimonies as much as I have done.

J. I. PACKER
Vancouver, Canada
Easter 2009

Acknowledgements

Thanks must go to all ten of these scientists who were willing to share their stories. It was a privilege to be able to carry out these interviews and edit the transcripts into the form you see here. My research assistant James Crocker was a huge help during every phase of this project. Several people read the manuscript at various stages and I am grateful for their feedback: Denis Alexander, Connie Bertka, Andrew Bowie, Mark Brickman, James Crocker, Jim Moulton and John Urquhart. Many others were involved, providing the information, advice and encouragement that are always needed in producing a book. Thanks also go to editor Alison Hull, and Robin Parry and the team at Paternoster for making the process of producing a book seem so easy. This project was supported by a grant from the John Templeton Foundation.

The Journey Begins

RUTH BANCEWICZ
*The Faraday Institute for Science and Religion,
St. Edmund's College, Cambridge, England*

Can science and faith be kept in separate compartments?

In some ways the central truths of Christianity stand apart from science but science does affect theology and the way that we live our lives: How can we use technology to care for the poor? What will you say when your doctor offers you a genetic test?

And some of the big questions in life demand answers that neither faith nor science can answer alone: How did we get here? How should we use the earth's resources? How do we deal with the fact that every new technology can be used for great harm as well as great good?

There are thousands of people who are equally at home in the church and the laboratory, and they have some fascinating stories to tell about their own discoveries in faith and science.

Being a Christian in Science

As a PhD student in a genetics lab at Edinburgh University, I used to find the list of senior scientists on my Christians in Science[1] membership information reassuring. This wasn't because I felt pressured in any way as a Christian working in science but sometimes I wondered how my career might fit in with my faith. It was good to know that there were people who had been there before me and had done extremely well.

When I finished my studies I worked as a research scientist for several months. The plan was to spend six months learning some new techniques before going to work in the US but I realized that research was not my niche. I found myself gravitating away from the lab bench and towards my desk and the writing, thinking and communication side of science. The opportunity came up to do all these things, minus the experiments, as the Development Officer for Christians in Science (CiS) and I grabbed it with both hands.

For three years I was in the happy position of being paid to interact with other scientists who were Christians. My work with CiS involved travelling around the UK, visiting universities, churches, conferences, Christian groups and the growing number of Christians in Science local branches. I spoke with a huge number of people about science and faith and heard all the different perspectives, especially on creation, which seems to be today's hot topic. It helped me to make far more sense of my own scientific and Christian knowledge than I ever could by myself.

Some people come to discussions of science and faith with an image of Christians drowning in a sea of scientific facts, unable to escape the inevitable death of religion. This view often comes from a Christian who's afraid or suspicious of science or an atheist who thinks science has killed God. In reality (and hopefully this is not a surprise) there are a huge number of scientists who are also Christians, and hundreds of books have been written explaining how faith and science fit together.

Surveys of the level of faith among scientists have produced some interesting results. The most comprehensive surveys have been carried out in the USA.[2] In 1916 the eminent American psychologist James H. Leuba found that only 42% of senior research scientists believed in a God who answers prayer. He predicted that religion was in a downward spiral and would gradually fade away, because science was "killing God."[3] People were shocked that scientific knowledge could have such a serious effect on society. In 1997 a second survey was published that shocked people in a different way. The lawyer and historian Edward Larson and author and journalist Larry Witham repeated the survey using exactly the same questions. They found that over 39% of scientists believed in a God who answers prayer, only a small reduction in 81 years, so the predicted downward spiral had never happened. It seems that the number of religious believers working in science simply reflects society as a whole,[4] rather than any specific relationship between science and faith. This is reflected in the make-up of any lab around the world today: you'll find Christians, Muslims, Hindus, Jews, atheists, agnostics and others working in all fields of science. They use the same experimental methods and present their results in the same way.

Interestingly, the number of religious believers among high-ranking US scientists (members of the US National Academy of Sciences[5] and those in elite universities[6]) is only 7%. The contributors to this book are a testimony to the fact that this is more likely to be a result of social and political trends in US institutions, including the separation of higher education into religious and secular domains, and not a sign that spiritual rot sets in when a scientist makes it to the top.

Discovering Christianity and Science

What I want to focus on in this book is how faith and science fit together in practice today. First of all, how do scientists find their faith? The scientists that I meet every day came to their Christian faith in different ways. Some were atheists or agnostics who looked at the evidence for the existence of God again as adults and changed their minds. Dr. Francis Collins, the former head of the Human Genome Project in the USA, and the scientist-turned-theologian Professor Alister McGrath, were both atheists who set out to find out about Christianity so they could shoot it down. They discovered something that changed their lives in a completely unexpected way. Others were introduced to Christianity at an early age and there was a point, often at university, when they decided to keep following or return to their faith. For example, Professor Sir Ghillean Prance, the former Director of Kew Gardens and former Director of the Institute for Economic Botany at the New York Botanical Gardens, was taken to church as a child but decided to become a Christian in his first week at university. For both these types of people, while they were in the middle of their scientific studies they discovered Christianity for themselves, showing that faith and science do quite naturally go hand in hand.

Science takes people into a huge range of careers: teaching, industry, writing, medicine, agriculture ... I've spent most of my time with Christians in Science interacting with scientists in a university setting. It's been interesting to ask what drew people into research in the first place. The main reasons seem to be a thirst for knowledge or wanting to make the world a better place: most scientists share both motivations but to different extents.

I did my PhD in genetics at Edinburgh University, based in the Medical Research Council (government funded) Human Genetics Unit. It was on the grounds of a hospital, so there was a constant reminder of what we were there for. When I walked through the link corridor to the hospital, that hospital smell would hit me, and I would see people being wheeled along the corridor. It was a bit of a shock sometimes, remembering that what I was doing was so far from actually contributing to a cure for anything.

I chose my own PhD topic partly because I knew it could have an impact on people, and not just rich people in the west. I was looking at environmental pollutants and how they affect unborn children. A developing embryo that carries a disease gene is sometimes completely healthy until an environmental disturbance like a toxic chemical triggers the onset of a (genetic) disease. I was looking at a specific biological "trigger mechanism" that was thought to be important at very low levels of exposure to environmental hazards. We found evidence that this mechanism does have an effect in fish[7] and, now that I have left research, it's up to others to look into whether the same effect also occurs in humans.

The scientists who are motivated by a thirst for knowledge often find themselves doing research that won't have an impact on people for many years. Sometimes they feel guilty that they're not directly feeding the hungry, healing the sick or in "full-time Christian work." A common thread through many people's stories is a slight feeling of guilt over having such an interesting job. Scientists often feel they should follow a more "spiritual" career. For example, Ghillean Prance was actually accepted for ordination in the Anglican Church before his future father-in-law encouraged him to keep on using his scientific talents. He finally realized that science could be just as important a calling as working in the church. We should be glad he did, because he was one of the first people to highlight the problems caused by cutting down vast areas of the Amazon rainforest. Many of the scientists that I meet have had to go through a process of learning to see their work as a vocation or calling, and finding their niche in science where they can worship God every day through what they do. Doing research where the outcomes are somewhat unknown is sometimes more difficult than working in a profession that will affect people very directly but is just as valuable and may end up affecting more people over time. It is very difficult to predict what benefits basic research might bring to humanity in the long-term.

So how does a Christian who is a scientist live out their life in the lab? People sometimes have an image of lonely scientists struggling through their research at all hours of the day and night. The all hours part is often true, but in reality most labs are stuffed with people like sardines, each with their own two feet of bench space. Sharing expensive equipment and taking turns to clean up are part of the enterprise. You'll find the whole range of different types of people in research: sorters, ideas people, visual people, perfectionists, technology lovers, mathematicians, philosophers, loners, socialites, and so on. People have to be curious, creative, able to communicate and, most importantly, persistent. Most scientists have something of an anarchist in them: they are always looking for ways to challenge old ideas, disprove someone's (or their own) cherished theory and build new ones. The average research scientist doesn't just do experiments and write papers. They also spend their time preparing talks, writing grant applications, teaching students, going to conferences and socializing with their colleagues. There are usually rituals and traditions that make up for all the late nights and weekends in the lab: firing champagne corks as far as you can down the corridor, welcoming new students over pizza, comedy (?) sketches at the department Christmas party and the essential celebrations when papers are published, grants come in or students pass their PhDs.

There will be those in the lab who are finding it hard to look beyond the daily round of experiments, trying to write their next paper and keep their funding. Some may be convinced that religion is irrational and that God is a distant or non-existent figure that we can't claim to know anything about. But the lab can also be an amazingly fruitful place spiritually, where you're surrounded by original thinkers who are faced with the beauty and complexity of creation every day. The Christians that I talk to often say that their faith develops as their scientific knowledge grows. In the following chapters you'll find a whole collection of stories that describe how studying the natural world can reveal more of God's creative power and invoke a sense of wonder, worship and awe.

Test of Faith

Several things took me by surprise when I started working for Christians in Science. One was the sheer enthusiasm of the people that I was contacting every day. They were keen to take any opportunity to speak or write about what they were doing and explain that, as Christians working in science, they felt like whole people, not torn between faith or science. Another was their level-headed graciousness when other Christians disagreed with them on an issue. Things can get heated when your cherished work or idea is debated. These people showed me how to leave behind my ego and approach things rationally and compassionately. I attended a conference where I felt that I had to take another look at an issue that I had already made up my mind about, which was a painful process. Afterwards I read John Stott's *The Contemporary Christian* and realized that I had been following his model

> I have myself tried to observe the rule never to engage in theological debate without first listening to the other person or reading what he or she has written or preferably both ... I am not claiming that this discipline is easy. Far from it. Listening with patient integrity to both sides of an argument can cause acute mental pain. For it involves the *interiorizing of the debate until one not only grasps but feels the strength of both positions.*[8]

It's when Christians and others learn to do this that the really interesting conversations happen. Even more interestingly, we can move beyond the differences to the exciting bits that we have in common. What are we doing here, and what can we do about the issues that we do understand more fully? When this happens, then we will really make progress: as scientists in the lab fully supported by the church community, and as people who want to engage with the issues and opportunities raised by science and faith.

In the course of my work I also found that pastors, teachers and others were crying out for a clear and accessible explanation of the interaction between science and faith for the "person in the pew" and for others outside the church community. It is often easier for academics to write scholarly publications than to communicate in an imaginative way with the general public. At the Faraday Institute I finally found myself in the position of having all the time and resources that I needed to focus on producing materials for the people who were demanding them, and I was able to bring together a team of talented and creative people to make it happen. This collection of life stories, and the DVD and other materials that accompany it, are the results of our work.

But why "Test of Faith"? The challenge that has been put forward so many times recently is that God is a delusion, and science has removed the need for faith in anything. How do believing scientists respond to this? They have all been trained to think and test ideas to the limit. If their faith and their science are both genuine searches for truth, we need to hear from them.

The aim of this book is to share the experiences of some Christians working in science with people who might be in the same position themselves, and with others who are drawn to look in on this world and see how it works. I imagine that most people who pick up this book will not be professional scientists but may have thought about science and faith, even if it was just to take in a news story or a review of the latest activities of the "new atheists."[9] As well as their personal stories, each scientist has chosen a short piece of writing or a Bible passage that has helped them to see their faith and science together in sharper focus. As you read the life stories of these scientist-believers, I hope that both their faith and their love of science shine clearly through.

1

Learning the Language of God

Francis Collins

Former Director of the National Human Genome Research Institute at the National Institutes of Health, Bethesda, Maryland

What if we could uncover the contents of the entire DNA instruction book inside every one of our cells, that drives the development and functioning of our bodies? This is the question that Francis Collins asked as the NIH Director of the Human Genome Project, a huge international effort that involved more than two thousand researchers. In 2000 the first draft of the DNA "genome" was completed after ten years of hard work. The official statement from the White House said that "We are gaining ever more awe for the complexity, the beauty and the wonder of God's most divine and sacred gift." This wasn't political spin for Collins but really reflected his own experience.

I grew up on a small farm with no plumbing, homeschooled by my mother and father until I was ten. I was given a great gift by my parents: the gift of learning to love learning and the discovery that new experiences could be some of the most exhilarating things that could happen to me. That gave me a sense of curiosity which worked its way through mathematics, chemistry and physics, onto biology and medicine, and then ultimately to the exploration of this amazing script called the human DNA genome.

My father was a professor of drama and my mother a playwright. We lived in this rather rustic environment, farming without any machinery, but they quickly realized they couldn't make a living that way. My father's full-time job of teaching was how he actually put bread on the table. My parents were very much in the theatrical mode and of course all their four sons were expected to be the same. I was on the stage by the time I was four years old and loved every minute of it.

Science was not something that was really part of my family experience. It became real to me at the hands of a charismatic chemistry teacher in a public high school in Virginia. He could write the same information on the blackboard with both hands simultaneously! More importantly, he taught us the joys of being able to use the tools of science to discover things we didn't already know. I caught that fever, and I've still got it.

At home faith was not something that was talked about very much. I wasn't really raised with any particular spiritual worldview. My parents were not people who criticized faith but they didn't consider it particularly relevant or important. I didn't see any evidence in my parents of their leanings in that direction, although ultimately my father did become a believer. I was sent to learn music at the local Episcopal church because they had a wonderful choirmaster and organist. My father made it clear that it wasn't really that important to pay attention to the sermons, so I learned a lot about music but I didn't learn much about theology.

I went on to study chemistry at the University of Virginia at the age of sixteen, because my homeschooling meant that I was two years ahead in high school. When those late-night discussions about religion began to occur in the dormitory I was skeptical about what the believers were saying, on the basis of their own upbringing, about the reality of their faith. Some of my neighbors were strong atheists who were, I thought, effective in their arguments. I found myself identifying with the skeptics and the atheists because I had no particular reason to attach value to a faith system. As a young man with lots of temptations, it was also convenient to reject the idea that I was responsible to anyone or anything other than myself. I slipped into what essentially was agnosticism (the idea that we can't know for certain whether there is a God or not), although frankly I didn't know the word at that point.

As a PhD student studying quantum mechanics[1] my passion was mathematics, and the way that you can describe the collision of atoms and molecules using mathematical equations. I believed that everything that happens in the world could be explained by reducing everything to this level, and that all our thoughts and actions are determined by these laws and equations. I was comfortable putting any religious beliefs down to superstition; the sort of thing that we should leave behind as we get more knowledgeable about how the universe works. I had no use for people who tried to argue that there was something outside of the physical world that was also valuable and true. I assumed that any religious feelings that anyone held must be because of some emotional experience (and I didn't trust those) or on the basis of some childhood indoctrination that I was glad to have missed.

In graduate school I decided I should broaden my horizons a little bit, and I took a course on biochemistry and molecular biology (the study of DNA). Until then I had not had much interest in biology or medicine at all. In high school I found biology boring, because it seemed to be largely about learning mindless facts. I had assumed it was just all murky and muddy and it wouldn't make any sense at all. The idea that there was this information molecule called "DNA" and that it was the way in which all living forms directed their material processes was truly exciting. I got the sense also that this field was breaking wide open and that there were going to be consequences for humans in terms of our ability to understand and perhaps treat disease. Combined with my worry that the most exciting discoveries in quantum mechanics had been made fifty years ago, this began to emerge in my mind as an alternative way that I might decide to spend my career.

Changing directions in a rather drastic way (I was already married and had a child at this point), I decided that I would go to medical school. And I found that I loved the experience of learning about the human body and all of its components. I particularly loved being introduced to genetics: DNA was mathematical in a certain way. But, later in my medical training, I found myself sitting at the bedside of patients with serious diseases. This was no longer an abstract study of molecules and organ systems. These were real people. I realized soon enough that the medical methods we had to help many of these people were imperfect and were not going to save them from death. Many of them had cancer, others heart disease – a variety of incurable illnesses. We could make them comfortable, and we might be able to slow down the disease for a bit, but ultimately they were going to lose their battle.

Up until then, the idea of life and death had been abstract for me but suddenly it was very real. I was puzzled by how these people in this hospital were, for the most part, not angry about their circumstances. I thought I would be. Instead they seemed to be at peace, realizing that their life was coming to an end. Many of them even talked about how their faith gave them comfort. This was the rock that they stood upon, and they were not afraid. I realized that I would be afraid. I didn't know what was on the other side; I suspected nothing at all.

One afternoon I was with one of my patients, a wonderful elderly woman who had very serious heart disease and had suffered mightily for it, and for whom we'd essentially run out of options. She had a particularly painful episode of chest pain while I was with her. She got through it, and then she explained to me how her faith was the thing that helped her in that situation. She realized that the doctors around her weren't giving her that much help but her faith was. After she had finished her own very personal description of that faith, she turned to me (I had been silent), looked at me quizzically, and said, "I've just shared my personal faith in Christ with you, doctor, and I thought you might actually say something but you haven't said anything. What do *you* believe?" Nobody had ever asked me that question so directly, and with such a generous, sincere spirit. I felt the color rising in my face, and I felt an intense disquiet about even being there. I stammered something about not being quite sure and left the room as quickly as I could.

Afterwards I puzzled over what had happened with that lady, and why it had been so unsettling. Ultimately I had to admit to myself that her question sought an answer to the most important issue that we humans ever deal with: is there a God? I had arrived at my own negative answer without ever really looking at the evidence – and I was supposed to be a scientist! If there's one thing that scientists claim they do, it is to arrive at conclusions based upon evidence, and I hadn't taken the trouble to do that. I was pretty sure there wasn't any evidence for God but I had to admit that I didn't know. I also had to admit that some of my teachers in the medical school were believers, and they didn't seem to be the sort of people that would stick to something just because they'd been told about it in childhood. I had wondered about that, and I'd never actually considered what they might describe as the basis for their faith. Maybe it was time to learn something about it? Maybe this wasn't all just superstition? Maybe there was at least something there to understand?

There are all sorts of ways that one might come face to face with this question of whether there is a God, but a particularly interesting one is sitting at the bedside of someone who is facing death and imagining yourself in that position. I couldn't help but think, "I don't want to be in that position and not have some better sense of the answer." When you're young you can imagine for the longest time that you're immortal, but as a medical student facing death every day on the wards, it was hard. That's what happened to me that afternoon: a combination of realizing I hadn't done the hard work that I should to answer a really important question, and a realization that my life was not going to go on forever. Thinking about that, there in my twenty-sixth year, sitting at the bedside of this wonderful, kindly, spiritual woman, I realized this was not something to put off.

That day at my patient's bedside started a journey for me, a journey that I was reluctant to begin but felt I needed to; a journey that I thought would result in strengthening my atheism. First I had to understand what religious people believe, and I had a hard time finding out the basic principles of the world's faiths. I was quite muddled about what they stood for. I went to a Methodist pastor who lived down the road and asked him about all this. He gave me a copy of C. S. Lewis's book *Mere Christianity* and told me that the author was an Oxford scholar, a prodigiously developed intellect, who had travelled the same path. Lewis had been an atheist, was puzzled by what his friends who were believers were talking about, and set out to disprove them. He found that the evidence went the other way, and ultimately became one of the most compelling Christian voices of the twentieth century. Within those pages I realized for the first time that one can come to a belief on a rational basis, and that in fact atheism is probably the least rational of all the choices.

It took me three or four months to get all the way through that book, because it was very unsettling to see that the foundations of my atheism were falling apart page by page and leaving me in a position of having to accept the idea of God's existence: something that I was not prepared for. I realized that atheism claims a "universal negative" (there is no God at all) which is a difficult thing to prove in any circumstance. I realized it was even more difficult given the many pointers to God in the universe: its beginning, and its fine-tuning in terms of the way in which all those physical constants that determine the behavior of matter and energy seem to have been set in a certain very precise range to make life possible. There were many other things, including my beloved mathematics and why it actually works anyway to describe the universe; something that makes you think the Creator must have been a mathematician. All of those things I found compelling, but they only got me as far as seeing the plausibility of belief in a deist[2] kind of Creator, a distant sort of God.

It was Lewis's argument about the moral law, this knowledge of right and wrong that distinguishes us from all other species, that I found most convincing and do to this day. It is a moral law that we break quite regularly but we know it's there. It often makes very little sense in naturalistic terms because it sometimes calls us to perform acts of radical self-sacrifice that are clearly not good for the passing on of our DNA, which is all that evolution by natural selection would care about. That part of the argument led me to acknowledge that if God exists, then God cares about people. Why else would this moral law be something that people, including me, experience? I began to realize that God was perhaps calling to me through a language I had lived with all my life without appreciating its source. If that was true, it also said that God is good and holy, and was calling me to be the same. Given all the times that the moral law had told me to do one thing and I had done the other, I was, and still am, hopelessly short of that.

The discovery that there might be a God who cared about me was a profound revelation, but I also began to sense a growing foreboding. I was beginning to discover God, but the character of this holy God was almost infinitely far away from what I might be able to approach with all my failings. That distress was blessedly answered as I began to understand the person of Jesus

Christ. I had thought that Christ was as much myth as history but I realized after reading more about him that he was a historical figure. There is a great deal of evidence for Jesus' existence and his teachings, and even strong support for his literally rising from the dead. This, while it seemed incredible at first, began to make the most perfect sense. I realized that I would be cut off for all time from God if I didn't have a bridge of some sort to make me right, given my imperfections and God's holiness. The perfect bridge, I realized, was Jesus himself. That was a joyous revelation but also a scary one. As it all began to fall into place, I realized I'd come so far down this road that it was going to be very hard to turn back.

In a muddle about all of this, on a beautiful afternoon (one of those rare moments as a medical resident where I had a little time off) I went hiking in the Cascade Mountains. It was a sunny day, the sky was perfectly blue, and I had that experience that we are occasionally given of being cleared of all of the distractions that otherwise get in the way of thinking about what really matters. I just left the car and walked up a hiking trail. I had no idea where I was, and it's a wonder I didn't get lost. As I walked up that trail I turned a corner and there was a sheer cliff face in front of me, at the top of which there must have been a small trickle of moisture. As that trickle came down the cliff it froze, and glinting in the sun was this frozen waterfall that came down in three cascades. I'd never seen anything like this before. It would take anybody's breath away, spiritual or not, to see this beauty of nature. But it caught me at a moment where I realized that this was an opportunity to ask the question that we all have to ask at some point. Do I believe in God? Am I ready to say yes to that question? And I found that all of my resistance fell away. Not in a way that I could tell you precisely, in terms of "Yes, I went through this logical argument and that theorem." No, it just was a sense of "I am ready to give myself to the love that God represents and that has reached out to me. I am ready to put aside my resistance and become the believer that I think God wants me to be." I fell on my knees and said, "This is something I want. Christ, come and be my Savior, and change my life. I can't do it by myself, and maybe tomorrow I'm going to think I was nuts but today this is real. This is the most real thing that's ever happened."

I was not quiet about my new faith. I was a young Christian full of excitement, wanting to share it with everybody. My colleagues were generally supportive, although a bit puzzled. A few of them, knowing that I was already on a pathway towards spending my professional career in the field of genetics, suggested that I was on a collision course and that my brain was in danger of exploding if I allowed my faith in Jesus and an exploration of genetics and evolution to come together. Those views would clearly be found incompatible and I would end up in some sort of misery and crisis.

But shortly after I became a Christian I realized there was no real conflict between belief in a Creator God and using science to understand how God had done that creating. It is well documented by a recent survey that 40% of scientists in the USA believe in a personal God. I can't imagine that science, which allows us to peer dimly into God's creation, would in some way threaten God. Here is an opportunity to understand God better and increase our awe for what God has created.

I have been more open in terms of talking about science and faith than many scientists have been. There wasn't much written about how to put these worldviews together, so I decided to speak and write more openly about it. This has, for the most part, been a really exhilarating experience, and has resulted in my having the chance to talk to thousands of people about a topic which often isn't discussed, and in a small way to encourage people to think these issues through and not just put them to the side. It's not necessarily an easy thing, though, for a scientist to talk about this. There's a bit of a taboo in academic circles about discussing matters of faith, and that topic will empty the seminar room about as quickly as any I know of. There's a sense that this is not what science is about, and that you should leave those conversations for your home or your church. I understand the reasons for that discomfort but I think it's unfortunate that this view has led many people to believe that science and faith are incompatible.

You can read the book of the Bible or you can read the book of nature, and you can find truth in both ways. You need to be careful, of course, about what kind of question you're asking, and which tools are appropriate for that question. It seems to me that to put either of those kinds of investigations off to the side and say "That's either inappropriate or dangerous" is to impoverish your opportunity to address the most important questions in life. We are only given a brief time to live here on this amazing planet, so why should we limit ourselves? We need to search in all kinds of directions for the truth.

Praise the source of faith and learning
who has sparked and stoked the mind
with a passion for discerning
how the world has been designed.
Let the sense of wonder flowing
from the wonders we survey
keep our faith forever growing
and renew our need to pray:

God of wisdom, we acknowledge
that our science and our art
and the breadth of human knowledge
only partial truth impart.
Far beyond our calculation
lies a depth we cannot sound
where your purpose for creation
and the pulse of life are found.

As two currents in a river
fight each other's undertow
till converging they deliver
one coherent steady flow,
blend, O God, our faith and learning
till they carve a single course
while they join as one returning
praise and thanks to you their source.

2

Being Human: More than a Brain

ALASDAIR COLES

Senior Lecturer in Clinical Neuroimmunology,
Cambridge University, Honorary Consultant Neurologist to
Addenbrooke's and Hinchingbrooke Hospitals,
and Curate at St. Andrew's Church, Cambridge, England

What are the most important issues for a neurologist? Alasdair Coles has encountered many questions in his role as a doctor and a researcher: Are we controlled by our brains? Is religion an illusion created by our minds? What does the study of brain diseases reveal about our personalities or our faith? How can scientists carry out research with human patients responsibly? As a minister, he has encountered other challenges: What is the place of the church in a hospital setting, and how does faith affect our career decisions?

Ever since I stopped wanting to fly fighter planes, at around the age of fifteen, I've wanted to be a neurologist. I specifically wanted to work with people who have diseases of the brain. The thing that has always interested me is finding out how much the physical structure of the brain can explain our behavior.

There was a specific event, or conversion experience if you like, that got me interested in neuroscience. I had one year left at school and was trying to decide what to do next. I went for selection to be a Royal Air Force (RAF) cadet, and I was successful in the preliminary stages. At the same time, a friend's father, who was a doctor, invited me to sit in on his clinic. I have a very clear recollection of what happened. A man of Asian origin came in who had great difficulty walking, with very droopy eyelids and slurred speech. Within a couple of minutes the doctor turned to me and the medical students I was sitting with, and said, "I know exactly what's wrong with this person and I know how to make him better." He gave him an injection and, as he did so, this man's eyes opened, his limbs became strong, he was able to talk clearly and he could walk. I thought, "That is fantastic, I really want to be part of this!" It was the combination of being able to diagnose something just by talking, listening and acute observation, then being able to test your diagnosis very rapidly and make someone better. It was powerful. I thought, "Actually, this is cooler than flying a fast jet."

I know now that the doctor was a neurologist, that the patient had an illness called *myasthenia gravis* and that, sadly, the injection he received would have worn off in a few minutes, but I continue to be impressed by how much you can do by simply talking to people and hearing what they've got to say. You can go a long way towards working out what's wrong with someone, even at a very detailed level, with just a knowledge of anatomy and the basic principles of neurology. That joy has stayed with me ever since and if I could chose how my future would go, I wouldn't want to ever drop that basic experience of going to a clinic and talking to people with brain diseases. The reason that I do research is because I'm not satisfied with just repeating what other people have thought or done. I'd like to make a contribution to what we hand on to the next generation of neurologists and to patients with illnesses now.

Finding Faith

I was brought up in the Middle East in a conventional secular family environment. I didn't have a lot of contact with the Christian faith until I was sent back to boarding school in the UK and experienced standard private school religion. I was an institutional success at most things, so I took part in institutional religion until I got to university in Oxford. While I was at medical school there I bought into the philosophy of Richard Dawkins's book, *The Selfish Gene*. He claims that our genes (DNA, the coded instructions inside every living thing) are what control our destiny, and that faith is simply a by-product of evolution that we don't need any more. I thought that was a satisfactory account of how things worked and had one or two debates with the Oxford University Christian Union on the subject.

For several years I was quite comfortable with all this. I thought Christians were slightly inadequate people. You didn't need a faith because Dawkins had it all sorted. Two things changed that view for me in a double-pronged assault. I began meeting a series of Christians who were clearly not inadequate and could not be written off, even by someone as arrogant as I was at the time. They were people who were very impressive in lots of different ways (and I eventually married one of them). The second major factor was reading C. S. Lewis's book *Mere Christianity*, which has affected so many people, especially logical-minded scientists.

The opening chapter of *Mere Christianity* hits hard in the place where Dawkins is weakest, which is the question of where goodness and self-sacrifice come from. Dawkins' account of the selfish gene isn't satisfying at all because when you look around at people, they do behave as if there is a good and a bad, and they do sometimes behave in unselfish ways. *Mere Christianity* takes that question right by the scruff of the neck and says, "This is a big problem. How do we account for our instinctive sense of right and wrong?" C. S. Lewis eventually led me to think that perhaps there was a God and perhaps that God was best understood as Jesus.

A year or two followed of feeling very destabilized, and quite unhappy with my situation. I didn't know quite what I believed in or what I could trust. I was spending a lot of time praying to a God that I wasn't sure existed, and struggling with a lot of the conventional language about God that was being used around me. I found it very hard to understand what people meant by "meeting Jesus," "having Jesus as a friend" or "encountering Jesus," and I'm very wary about using that language now because it can seem so odd. I now understand it as a kind of shorthand but at the time I thought, "This is nothing like what I am experiencing. I must have missed something."

I had been confronted on the rational front by C. S. Lewis and on the personal front by encountering Christians whom I respected. The final event was a spiritual experience when I was in Chile during the last year of my medical degree. I took a Bible with me and read it every day. I often prayed but I was still not quite sure what I was doing.

I decided to go away for the weekend to a huge national park in the center of Chile: a vast expanse of uninhabited land full of lakes and volcanoes. As I went through the gates I learned that there were only two other people in the park that weekend. I started to walk around one of the lakes but the way became increasingly difficult. First there were little streams to cross over, then there was a bit of a climb and then the path seemed to disappear. There was one place where I had to scramble across a cliff face in such a way that I knew I couldn't go back. My fingers were starting to bleed and I kept slipping down. At one point I had to swim and it was very cold. I was getting more and more frightened and I remember praying, "Please save me from this." It was about then that I noticed two people in a canoe in the middle of the lake. I fell and hit my head and this canoe just kept drifting by. I shouted to them but they stayed about two hundred yards away, just watching me. When I finally got round the lake several hours later, I was in a terrible state. The two in the canoe, a man and a woman, had built a fire and cooked some food for me. When I asked why they hadn't picked me up, they said they weren't going to rescue me but were just there to look after me. The man explained that he had once had cancer. His radiotherapy meant that they couldn't have children together but at least he had survived. They came to this lake every year to remember what they had gone through. He said, "For a moment you were our child today" and I was really struck by that.

To this day I would say that my time at the lake was one of those events that takes on a supernatural quality. As I thought it over, I realized that God was looking after me in a special way through those people. He wasn't rushing to help me, pick me up and sort me out but he was there. That's served me very well as a picture of what I was going through at that time. I was destabilized and struggling but there was a God waiting nearby. Having that experience, and the interpretation of that experience, felt like a gift of faith. I became a Christian at that point and I haven't really struggled with my faith since.

Impacting Lives

One of the very powerful things for me and where I learn about God, is seeing the way that some people with a neurological disability are cared for. I am often brought to the point of tears by encountering patients who were born with serious disability, often with epilepsy and brain damage, who are now fifty or sixty years old. Their parents, now aged about eighty, are still attending to their children's daily needs, and often their main concern is what's going to happen when they die. These are real heroes, spiritual giants who are showing God's love in tiny little pockets of isolated existence that go unnoticed and without any kind of praise.

Watching people face neurological deterioration bravely is another powerful testimony. One of the patients I've seen recently, for instance, is a young fighter pilot who has motor neurone disease and can only move his neck. He's gone from flying a top-class machine to being in a wheelchair, but in the last few months he's fathered a baby and he's looking forward to the future, whatever that takes. That is a very powerful image for me: someone showing an amazing capacity for being human in the face of a body that's falling apart. When I see people like this it is absolutely obvious to me that we are more than just our bodies. The only way I can understand it is through the Christian story of God breathing a spirit into us.

In contrast to some scientists, the impact of our research is immediate and can be devastating. We have a hypothesis about the cause of a disease and we give treatments that we think make sense. Those treatments have risks associated with them. The experimental drugs that we use to treat multiple sclerosis are powerful and some of the risks we can predict but some we can't. In a project that is designed to do good, some people will suffer rare side effects which cause harm. We have come to the conclusion that the treatments we use are likely to be most effective when given early on in the disease process. So we are faced with the dilemma of giving experimental drugs to young adults who appear entirely well, usually in their early twenties, with hopes and desires for the future, family and career. They are faced with the prospect of a disease which has progressive disability as part of its possible symptoms. Taking these drugs may help them but may also have significant side effects. So when things go wrong in our clinical trials it always affects a specific person who we know and see regularly.

In conventional terms the research that I do is all "ethical" because an ethical committee has reviewed it but actually I don't think it's enough to say that. I don't think it's enough to hide behind paperwork and committee activity. It's our team who have developed the strategy for giving a particular drug to particular people and are responsible for the harm that it does as well as the good. We are intervening in people's lives and their diseases; we're not just passive observers. This is where doing science is at its hardest and it asks questions of my faith. How reasonable is it for me to intervene in the lives of other people? How reasonable is it for me not to intervene when they are faced with a terrible disease? Do I give these experimental drugs as part of collaborating with God in God's work to fight disease or am I somehow pretending to be God and influencing things? These are challenges for me, and they're very painful human challenges.

Life Choices

When I got back from my trip to Chile I went through a short time of saying, "Now I'm a Christian I ought to do 'worthy' things." I briefly explored being a missionary doctor in Nigeria. I was there for just three months with my wife and (medically) it was a disaster. The skills that I have are not the skills of a mission doctor and that was a hard thing to learn. Looking back, I don't think God was prompting me to go to Nigeria at all – but that was my own limited vision of what a Christian could be.

I wouldn't say that nobody should go abroad to be a doctor. More recently I went on a trip to India that played to my strengths. I went there to raise awareness of multiple sclerosis, which is not well recognized in India, and I have the authority, the credibility and the skills to do that. I'm very open to going abroad again and working for a mission organization, but only in a context where I'm truly fulfilling my vocation and my skills. I believe that we all have a calling or purpose, the Bible seems clear about that. It's important that through prayer and by talking with those around us that we find God's purpose for us and are obedient to that.

It's difficult, particularly for someone like me who's in a very structured medical hierarchy with a clear career progression, to actually give up control. One of the things that has happened to me over the past few years is that I've gone into a period of being destabilized again. My career aspirations were conventional: to become a professor and head of department. That was what my goal was, but over the last few years I've lost that and I don't know what God wants from me. As part of that destabilizing there has been a feeling that I should learn more about the Bible, learn about the legacy of generations of Christians before me, and get some accreditation as a minister. I think part of it is to have a Christian voice in the workplace.

Spirituality and the Soul

In a church environment my interest in the brain and its behavior is obviously out of the ordinary compared to a lot of ministers. I often talk about my amazement at the way the brain is put together, and in a pastoral context I'm much more open to the body influencing our state of mind. One of the attractions of the Christian faith for me is that it's "embodied." The Bible recognizes that we are bodies as well as rational moral beings.

We are dependent on our brains to be conscious and aware, to reflect on ourselves, to have moral reasoning and to have desires and hopes. We have a brain that has the capacity for religious experience. We may or may not find ourselves in an environment (social or biological) which allows this religious expression to thrive. Each person then has the ability to decide for themselves whether or not to allow this spiritual awareness to flourish and at that point God can intervene. When you deal with someone who's struggling, looking at their physical state is absolutely as important as their spiritual and mental state. I think that is a very important first step for tackling problems: staying fit, getting rest, not drinking too much, not abusing drugs. You've got to make the best of that person's brain and body and then they're giving themselves the best chance to be spiritually aware.

I believe it is a mistake to say the "soul" is an independent entity that tells the brain what to think. It is also a mistake to say that the soul is nothing but the brain and that everything a person does is explained by the neurons (brain cells) themselves. It's obvious to me that that isn't true or at least not true in any helpful ordinary way. Consider how do we best explain personality? How do we explain human behavior? You could explain a painting on the basis of the chemicals that make up the oil paints. That would be a perfectly true explanation but most people would find that thoroughly unsatisfactory as an explanation of a painting. They would want to say, "There's more to it than that, someone made this, it has a significance that is dependent on the oils but ultimately it has nothing to do with them." Human thoughts and behavior are dependent upon but in some way separate from the material of the brain. You can either believe that God made us or that evolution made us or both, but we are made for a purpose.

Faith and Work

I think people are positive within the church about having someone teaching and preaching who has a life outside of the church. There can sometimes be a concern that the leadership comes from within an environment that is out of touch with the rest of the world. One of the things that has always irritated me about the churches I've been in is that it's possible to feel valued only for what I do for the church: for the coffee rota or for the efforts I make for the youth group. It's a very limited perspective of what the life of a church is. People who have a job outside of the church are increasingly important in the way our churches work and role models are important.

My hospital community lacks prominent Christian voices. It's interesting that the hospital chaplains will tell you that the group of people they have most difficulty approaching are the "alpha male" senior doctors. We are a very distinct tribe and closed off. Academic neurologists are a very unspiritual group of people. It is very unusual for a neurologist to be an accredited minister, and religion and spirituality are not welcomed as topics of conversation. I've never encountered any hostility. I've certainly met with curiosity but rarely positive support. The most common reaction is lack of interest or a feeling that this is slightly eccentric. However, one person has made a great deal of difference to my faith in the workplace. I am very fortunate to have a friend and work colleague who is a strong Christian. We agreed a few years ago to meet together to read the Bible and pray once a week. Then we decided to open up to all Christians. Now up to fifteen people meet once a week to study a Bible passage and pray. We pray for the hospital, for the people working here and for the patients. Leaving aside what that means for the institution and whether there should be more of it, for me it's powerful that I bring my faith to work and that people around me know that I'm a Christian so I can be held accountable for that.

Understanding the Mind

The "added value" of having a faith comes in lots of different ways. One of the things it has done is to make me ask if there is a neurological basis for religious experience. How do we fit faith into the working brain and at what level? These issues have never been a problem for my own beliefs. I think if anything my faith has encouraged me to keep asking questions, because at heart I think I'm just a child who's enthralled with things. I have come to understand that feeling of pleasure and joy as a gift from God and an encouragement for me to carry on.

My interest in the structure of the brain and how it affects behavior is stronger now than it ever has been. I would say that my faith encourages me to look into "the book of life" and read the work of God. Neurology is what I like and what I'm good at, and I think God shares that pleasure with me. It's never been a problem for me and I'm always surprised when I meet people who talk about conflict between the two.

When I read about God "knitting us together in our mother's womb" in Psalm 139 I just can't help having a picture of God cheerfully knitting together the highly complex pathways of the brain, interweaving this structure with that, chuckling to himself at how puzzled humans will be about his creation. I know, of course, that this is completely unlike our understanding of the biology of the developing brain. Yet the pictures in that psalm resonate as richly true nonetheless: we are put together in a secret place, thoroughly organic and embodied, yet at the same time "hemmed in," protected and limited by a God who knows us fully and personally.

Psalm 139:1–18

O LORD, you have searched me and you know me.
You know when I sit and when I rise; you perceive my
    thoughts from afar.

You discern my going out and my lying down; you are
    familiar with all my ways.
Before a word is on my tongue you know it completely,
    O LORD.

You hem me in – behind and before; you have laid your
    hand upon me.
Such knowledge is too wonderful for me, too lofty for me to
    attain.

Where can I go from your Spirit? Where can I flee from your
    presence? If I go up to the heavens, you are there; if I
    make my bed in the depths, you are there.

If I rise on the wings of the dawn, if I settle on the far side of
    the sea, even there your hand will guide me, your right
    hand will hold me fast.

If I say, "Surely the darkness will hide me and the light
    become night around me," even the darkness will not be

dark to you; the night will shine like the day, for darkness is as light to you.

For you created my inmost being; you knit me together in my mother's womb.
I praise you because I am fearfully and wonderfully made; your works are wonderful, I know that full well.

My frame was not hidden from you when I was made in the secret place. When I was woven together in the depths of the earth, your eyes saw my unformed body. All the days ordained for me were written in your book before one of them came to be.

How precious to me are your thoughts, O God! How vast is the sum of them!
Were I to count them, they would outnumber the grains of sand. When I awake, I am still with you.

3

Exploring God's Universe

JENNIFER WISEMAN

Astrophysicist

The universe is an incomprehensibly large and beautiful place. Looking up at the night sky has moved Christians throughout history to worship the God who made it all. Are there any challenges to faith when we look more closely? How far should we go in exploring it? Can we justify doing multi-billion dollar research when there are so many more immediate problems in the world? Jennifer Wiseman has a unique perspective on these issues, having spent her career working on research in astronomy, and then at a government level on public policy and national science oversight, before returning to lead an astrophysics research group at NASA's Goddard Space Flight Center, one of the world's best known space research centers.[1]

From Cows to Comets

My life began quite far from the formal world of science. I grew up on a small family cattle ranch in Arkansas. My parents did not go to college (they didn't have the means to do that) but they were very supportive of their children getting higher education, so we all did. One thing that got me interested in space was just looking up at the sky and being amazed by it. We would go on a walk every evening up and down the country lane that leads into the farm. Far away from any city lights, it was so dark that you could see a huge number of stars. In fact, I've never had that experience since, because in all my formal education in astronomy I've been in cities where the lights drown out the stars in the night sky. It's ironic that I could appreciate the heavens much more as a child in the country than when I've studied them formally.

The other big driver for me was about the time when I was in school. There was a series on television called *Cosmos* that featured Carl Sagan, a famous astronomer. He talked about how probes that NASA had launched into space were going out to Jupiter and Saturn, taking pictures of these planets and their moons and sending them back to Earth. You could see new worlds in close-up imagery: volcanoes on Io, this world of fire and activity, and cracked ice on the surface of Europa. It was these fascinating things that made me want more than anything to be able to know about what it would be like to explore the surfaces of these moons and other planets. I thought at that time, and I still believe, that surely one of the greatest things that humans have ever achieved is to send those probes out into the solar system to take pictures and send them back to us so we could know what's there, just exploring. I loved the idea of exploring space.

I liked all my subjects in school but I was pretty good in math and science, so my teachers encouraged me to go on. My oldest brother, who was the pioneer in the family, had already gone on to study in graduate school. He earned a PhD in economics and he gave me recommendations about where to try to go to study science, so with the encouragement of teachers and family I was able to go on and study science in university.

It's true that in his programs Carl Sagan did not have a lot of affirming things to say about the Christian faith that I grew up with, but I was much more taken by the positive aspects of his program. There was this whole universe of exploration and I was just enamored by what was out there that we had not yet discovered. I was already a Christian as a child. I had been raised in a Christian family and in a loving church. In our faith tradition you were baptized when you were old enough to make that decision, so I had been baptized at age ten or eleven and was sincere about it. I had experienced what I believe is the love of Christ and seen the genuine love and presence of Christ in other people. So there wasn't anything that I could be presented with in my science classes that would have put any doubt in my mind about God. It was an outpouring of my faith in God and my amazement at creation that made me want to learn more and more about what was out there. Since then my Christian faith has been a matter of learning how to follow Christ in very different settings from where I grew up: What does it mean to be a Christian in a scientific setting, on Capitol Hill or at NASA?

I was never afraid that my faith would be threatened by going to a secular university. In fact I think I had a prejudice that was almost the opposite. I was afraid that if I went to a strictly Christian college that somehow I wouldn't be hearing everything. I wanted to go to a secular university so that I could hear many points of view, and so I was almost biased the other way.

At university I knew I was interested in space but I didn't know whether I wanted to be an engineer building these probes that went out to take pictures of other planets or an astronomer that uses telescopes and studies theories of the cosmos. As an undergraduate student at the Massachusetts Institute of Technology (MIT)[2] I chose to major in physics, basically because I didn't know what I wanted to major in and I knew that the study of physics would still keep my options open. Physics is a science that you can apply to many different fields of science and engineering. Towards the end of my undergraduate years, I took an astronomy class that I found so interesting I decided astronomy was what I wanted to do. For my undergraduate thesis I studied a comet (114P/Wiseman-Skiff) that I unexpectedly discovered on a field research project which was exciting. I applied to graduate schools in astronomy at that point and ended up going to Harvard.

I did the research for my PhD on how stars form. Many people don't realize that stars are still actively forming in the interstellar gas that is all over the place in the galaxy between stars. So I studied how stars form, using these big dish radio telescopes that receive light that is of a different wavelength than the light or colors that our eyes are sensitive to. It's a very interesting process: the interstellar clouds of gas become unstable and start collapsing into dense clumps of gas that then coalesce into new stars. Studying that whole progression is quite fascinating. We learn by observing different regions that are at different stages of this development. The telescopes I used were in New Mexico: the Very Large Array (VLA) of radio telescopes that were featured in the movie *Contact*, although we don't actually use them to listen for alien life ...

Integrating Faith and Science

I was anxious to find other Christians wherever I went to study. I was very grateful in graduate school that we had a strong group of Christian students and there were some staff workers with the group there – you might call them chaplains. They were very influential in introducing me to the idea of doing scientific research as a means of serving God and as a means of pleasing God in and of itself, not as a separate thing. In other words, you don't study about God in church and Bible study one day and then think only about your science on a different day. These are not necessarily separate activities. God is just as interested in my understanding Scripture as he is in my ability to understand the science that I'm doing, and to relate these discoveries to other people is an act of service. Just being inspired that graduate study and beyond, which can be very difficult and discouraging sometimes, can actually be seen as a calling of God that is as valid as any other kind of Christian service, was a great encouragement to me.

My husband Mark was another great encouragement. We didn't get married until we finished our education but I met him when I was still an undergraduate. At that time I was wondering, "Should there be some kind of conflict in my mind between science and Christian faith?" I had heard some Christians talk very skeptically about the discoveries of science and I wondered if I needed to take on that same skepticism or feel some sense of threat or even hostility towards some of the discoveries of science. I just wasn't sure what to think about it. Mark first of all surprised me by telling me that the evidence for many of our scientific discoveries is in fact overwhelmingly convincing, so I should not only appreciate that and not feel threatened by it but also rejoice in it.

Since then it's been my experience that there's not much overt hostility toward faith, or even in particular Christian faith, in the professional field of astronomy. I think the reason is because there's actually not much theological conflict between the Christian idea that there's an ordered universe that had a beginning and our current scientific understanding of the universe.

The things we're discovering as scientists do blow our minds in many ways and change what we may have thought traditionally about where we came from or the age of the universe; yet this is something not to feel threatened by but to be very excited about. These are new treasures that God is allowing us to discover. These are secrets that God has known throughout time and only now have we been given the tools and the capability of understanding more and more about creation, and more and more about the universe. We shouldn't be threatened by it or hostile towards it but in fact take ownership of it. This is God's creation and Christians of all people should be the most excited about it, if we really believe that God in Christ is Lord of All – all time and all space. I think that idea first startled me and then set me free to enjoy the discoveries of science, even if we don't know all the answers as to how it all fits together.

If someone asked, "Has your faith directly affected how you take data with a telescope and interpret what you're seeing scientifically?" the answer to that would be, "No." That's a good thing, because science wouldn't work well if people looked at exactly the same data and came up with very different conclusions on objective results because they're coming from different philosophical backgrounds. For example, gravity and radiation should provide the same observed results to everyone, when accurately and carefully measured! That being said, I believe that my faith has had a big impact on how and why I do my scientific work more generally. I believe God has directly guided me into fruitful choices of research and into being able to help my co-workers along the way. God has helped me by providing wisdom and guidance when I needed it on choosing a career path or making choices about where to work. God has helped me not to fret as much over many things that would perhaps lead other people to be very worried about whether they'll get that next job or be able to have a fruitful career for the rest of their life. Christians are told not to worry about our lives but to trust each day and each step to God, so in terms of a personal faith in God I've felt a very real presence of God guiding and helping me throughout my scientific career, and I'm grateful for that. Also, because of the faith God has given me, I can rejoice in his power and creativity when I see, for example, the magnificent and beautiful galaxies and nebulae we observe in space.

Prayer

I don't think that the physical world of material things and forces is all there is. I believe that there is an all-powerful God who is responsible for the universe, and that there is a spiritual realm that cannot be measured with our tools of science. It's in that spiritual realm that God is also active. Christians believe that God is present both as "God the Creator" and also as "God the Holy Spirit," who is the present way that God speaks to us and guides us. It's perfectly natural then to be able to converse with the God of the Bible who is a personal God: this God knows us individually and encourages us to communicate through prayer.

You may legitimately ask whether it's appropriate as a scientist to pray for something that would change part of the physical world. In classical physics there's a cause and effect mechanism, so it doesn't make a lot of sense to drop a rock from the top of a building and then pray and ask God to stop it from hitting the ground. You're asking God to violate a physical law that he has put in place. There are two ways of addressing that. One is by going into the realms of chaos theory and quantum mechanics[3] which allow for uncertainty in the physical realm, so there is room for unpredictability and the idea that not everything is pre-determined from the beginning of time. There's a lot of uncertainty built into the system, so that gives some freedom for prayer to be answered and for God to work. And then perhaps there are also other dimensions in which God is working. That's the way some scientists address this.

I would add quite simply that the God of the Bible is a God that rules over the physical universe, and so this God is in fact capable of suspending or changing time and the physical laws as he sees fit at any moment. God could certainly work through the physical universe in the realms of other dimensions or chaos theory or the unpredictability of quantum mechanics, because God is responsible for all of those aspects of the physical world. Or God can simply act as he chooses, because God is sovereign over the physical world, and that is perfectly consistent with Christian faith. So I do believe that you can pray and supernatural things can happen. I also believe that prayer changes us so that, although sometimes God will answer prayers for healing, he also works in us to help us see things in a different way than we would have otherwise. These are workings that are of a spiritual nature that cannot be measured scientifically.

I don't think it's true that scientists necessarily struggle with needing physical evidence for their faith. I believe that's true for some scientists but not all. I think it depends on different personality types and the paths by which people come to faith. For me the harder issues are the same things that all people struggle with. Why would God allow a universe and particularly an Earth to exist where there is intense suffering and injustice, not only toward humans but to the rest of creation as well? Why doesn't God just intervene and stop innocent children (or even animals) from suffering because of what we do to them or to their environment? Why don't we use the scientific knowledge that we gain in one field to inform another field? Why do diseases persist that kill countless children before they reach the age of five, when most of these diseases could be prevented? And the concerns extend beyond our human realm. For example, we now know the needs of animals for their habitat and their psychological and social needs better than ever before, and yet we don't apply that same knowledge in our scientific study and treatment of livestock in agriculture. I believe that for Christians all of this should be overarched with an understanding of ethics and compassion and a worldview that encompasses a broader range than our immediate needs. I don't understand why people, even Christians, seem rather blind and callous at times and I wonder why God allows a lot of our inhuman actions to go on. But these are questions that Christians have asked throughout time ...

A Change of Track: National Science Policy

I've had a very interesting and perhaps non-traditional career path since university. When I finally finished my PhD, I went on to do research with the organization that runs the VLA: the National Radio Astronomy Observatory. I worked for three years in Virginia doing research using the VLA and other types of telescopes to study the process of star formation. I then went on to a second research position as a "Hubble Fellow," working on star formation at Johns Hopkins University in Baltimore.

As I went into the second research position, I began to have a stronger interest in the bigger questions, the big picture of why we are doing science. Why does the USA choose to fund certain types of science over others? How do we decide what the priorities are? Then as a Christian, I wondered: how do we bring together our religious understanding of God and God's plan for people, with this scientific study of the magnificent universe, with a cosmos that's been growing and changing since the beginning of time, and supports life on at least one planet? How do we bring that all together? So I did a lot of praying and soul searching about whether I should go on the traditional path which would have brought me to a university faculty position or try something different. I was offered a faculty position but at that point I was also offered the opportunity to become what's called a "Congressional Science Fellow." This is a one- or two-year position in Washington, DC working with Congress to help advise them on science issues. This was an amazing opportunity and I accepted it. It was a difficult decision but I felt that I wanted to understand the bigger picture of why and how our nation supports science. I got a lot of advice and spent a lot of time thinking and praying about this decision. Once I had decided to follow this more adventurous path I never looked back: it was a very exciting way to go.

It was a wonderful experience working with members of Congress and their staff, and helping them understand the scientific issues that are facing our nation. I worked with the science committee of the House of Representatives (part of Congress). In that post I interacted with federal agencies such as the National Science Foundation and the National Aeronautics and Space Administration (NASA), learning a lot and then helping our committee staff understand space science issues in these organizations that we oversaw for Congress. After that, since I understood these federal agencies a little bit better, I had the chance to work for the headquarters of NASA. I worked there for three years as the "program scientist" for the Hubble Space Telescope, overseeing the science mission of this satellite in space that continues to take marvellous pictures from its orbit, way above the clouds.

I still work with NASA but now I'm leading a research group at NASA's Goddard Space Flight Center. We are a group of scientists who are interested in everything about stars: we call it stellar astrophysics. We are also very interested in "exoplanets," which is a relatively new and "hot" field of learning how to detect and characterize planets that are orbiting other stars (i.e., stars other than our own sun).

Why Explore Space?

I think in many career paths you need to reflect every once in a while and ask, "Is what I'm doing worthwhile? Given that there are such terrible problems in the world, if I'm not directly addressing one of them, is it worthwhile for me to be doing anything else?" The answer for me has always come to this: I believe God gives us different talents and interests for a purpose, and we are to use our talents in different ways. I believe God gives some people keen interests in art or science for the purpose of uplifting the human spirit through discovery and exploration. I don't think that God created an entire universe and then wants us to ignore it until we figure out all our problems here on Earth. If you look at the very beginning of our biblical record, one of the first things that God asked the first human to do was to look at all the animals and name them. There is a sense of God saying, "Look at what I've created! I want you to go out and be as excited about it as I am, and to explore and understand these different creatures that I've made!" I think we can take that story and expand upon it. Christians believe that God is responsible for the entire universe, so wouldn't he be disappointed if we didn't even have the curiosity to go and learn? I believe that scientific exploration is a way of glorifying God by exploring what he has created.

We're told not to be overcome by evil but to overcome evil with good. I think there are at least two ways that we can do that in this world and we need both. One is to directly address the suffering of humans and all creation: stopping war, protecting the fragile, bringing food to the hungry, and bringing justice, compassion and freedom to the oppressed. The other is to provide alternative ways to focus our human energy and talents that are positive, that bring joy, and bring praise to God. One of those ways is science and scientific exploration. We need to conduct our science ethically and to use our scientific and technological advances for good and uplifting purposes. And we need to make sure that we share our discoveries with people of the world who may not otherwise get a chance to share in the joy. We are all part of a majestic and magnificent universe just waiting to be explored, perhaps mostly with telescopes, minds of curiosity, and hearts of praise.

Psalm 96:1–5

Sing to the LORD a new song;
sing to the LORD, all the earth.

Sing to the LORD, praise his name;
proclaim his salvation day after day.

Declare his glory among the nations,
his marvellous deeds among all peoples.

For great is the LORD and most worthy of praise;
he is to be feared above all gods.

For all the gods of the nations are idols,
but the LORD made the heavens.

I like this passage, because it not only affirms that the Lord is the only true God but it also gives a succinct but truly cosmic piece of evidence as to why there is simply no comparison between the Lord and all other "gods." Instead of trying to describe all the attributes and "marvellous deeds" of this praiseworthy God, the psalmist simply gives us a clue to God's unique character by pointing to what God has made. The heavens are beautifully ordered, and mind-boggling in size and content and activity. Every galaxy, star and planet has its own story, its own history, its own place in the ever-evolving universe. And within this universe we ourselves have been given the grace to live, to breathe, to discern good and evil, and to observe God's marvellous handiwork. What a joy, what a relief and even sobering realignment to remember that, come what may, we are known and loved by a living God who is responsible for the entire universe, and who also knows and calls us each personally! What "idol" or worry or distraction or goal can compare to this true God? Nothing.

4

Biology, Beliefs and Values

John Bryant
Professor Emeritus of Cell and Molecular Biology, Exeter University

The natural world is incredibly beautiful, and biologists are often awestruck by the intricacy and complexity of the living systems that they study. Research in this area has brought us huge advances in agriculture and medical science. How can scientists be involved in directing the development and use of this technology so that it is not abused? Do Christians have anything unique to say in this arena? Is science really accessible only to an elite few? John Bryant has had a lifelong interest in biology and has shared that with students of all kinds on both sides of the Atlantic. His interest in "bioethics" has grown into a commitment to teach, write and speak on the different challenges we face in using new technologies well.

My story starts with a working-class Christian family living in a council house[1] in south London. We had no car, so we did a lot of walking. We lived near the edge of town which meant that within minutes we could be in open parkland, woods, farms or open hills. In a sense, your horizon was as far as you could walk. At elementary school I had an enthusiastic teacher who was very supportive of my interest in the natural world. That combination of things meant that I grew up with a passionate love for nature and a knowledge that it was created by God – in the way that a small boy would understand that.

One of the people who has influenced me most in life is my dad. People find it extraordinary that a man who left school at fifteen could be such a major role model but everybody tells me, "You're just like your dad." They don't mean in looks, although now that I'm older, when I look in the mirror I see my dad. In attitudes he

was an extraordinary person. He cared about people. He believed
in the equality of the sexes and would share the housework. As a
postman he did his job to the very best of his ability, because he
was serving God in it. He was a working-class Christian man, a
caring gentleman and a brilliant father.

Faith was assumed as part of my family background, and in
my young teens or maybe earlier, it's difficult to say, it became
my faith too. The council estate where we lived was quite new,
so there were no churches in the area. There was a Sunday school
and an evening church in the local school hall, to which we
went. It wasn't until I moved to a different school that I really
encountered any Christian youth groups. My faith was nurtured
both in a Christian Union at that school, and by going along to a
Christian youth club called "Crusaders."[2] When you've grown
up in a Christian home your faith will have dips; it's inevitable
because you wonder whether you've just followed your parents.
I had to go through periods of evaluation as an older teenager
and even as a student.

I won a scholarship to the local independent school which was
a great privilege and changed my life. My love of the natural
world turned into a degree in natural sciences, and then a PhD
in biochemistry. I studied at Cambridge University, which was a
place for enquiry and debate. I was a student in the sixties when
everything was in a state of ferment, and so everything was up
for discussion. The Christian Union put on some science and faith
lectures which attracted the big names, so it was not unusual to
find a crowd of five hundred filling a lecture hall. I saw Francis
Crick (of "Watson and Crick," the two who discovered the DNA
double helix) at one such lecture for example. He was very anti-
religion, and I think that he possibly turned up to that particular
lecture, just to see whether he could shoot the speaker down!

I had been quite apprehensive about going to Cambridge.
Even though I'd had a few years at an independent school I was
afraid that people would be much brighter than me. I was also
anxious that they would find arguments to demolish my faith, but
that didn't happen. That kind of environment strengthens your
faith because you have to think more. By the time I was about
halfway through the first year I realized that actually I was okay
there. Having stopped cross-country running, because I thought

I wouldn't have time to do it, I started again and went forward in the running world as well.

The relationship between science and faith never worried me. I have never seen that there's any kind of clash between them. There were things that I needed to think about and work out, and there are things that we will never know the answer to, but it's never worried me. I find it very odd that some people on both sides of the debate think that they are incompatible. I think I was fortunate because I had some quite thoughtful science teachers in my high school, two of whom were practicing Christians. Although they weren't overt about their faith, I learned about it.

Ethics

As a young scientist I learned how to work with DNA, the code inside each cell that drives the growth and development of every living thing. At that stage I was working on the mechanisms that cause certain parts of a plant to grow at specific times. It struck me even then, in the late sixties and early seventies, that as we learned more and more about DNA we might be able to start manipulating it (genetic modification). My PhD supervisor once said, when he learned that my future mother-in-law had typed my first year report, "What did she think about this, John, you messing about with the very stuff of life?" It hadn't struck me like that before but from that moment onwards it did start me thinking. A couple of years later, when I was at the University of East Anglia, the local Christian graduates' fellowship asked me to talk to them about where work on DNA was going, and we had a good discussion. That was in 1970, and from that moment onwards I've had an interest in the ethics of biological and biomedical science, or "bioethics."

I think God has given us curiosity about the whole world. The search for knowledge is a worthy and worthwhile activity, and I don't think that any areas of knowledge are "outside the fence." There are methods and applications which I think do lie outside the fence for society in general, and for people of faith in particular. For example, I don't think we should ever contemplate experiments on human genetics that involve trying to get people to breed with each other (and this has happened in the past). Applications of molecular biology (the study of DNA and associated molecules) in biological weapons, and encouraging things like smoking, are off limits for me. There are many other areas of science that have a mixture of potential benefits and dangers, such as the use of different human tissues or making genetically modified (GM) organisms. We need to think about how to direct this type of research and the development of new technologies in responsible ways.

My involvement in bioethics really took off when GM crops became controversial in the 1990s. I was encouraged to think about this more deeply by my friend Suzi Leather, who later chaired the committee that regulated all the research on and use of embryos in the UK.[3] At that time she was in the same church as me and we spent a lot of time talking about GM crops. We started teaching bioethics to biology students, firstly through seminars and then through a major module. That module has now become embedded as a core course in the biology degree programs at Exeter University. It is also being used as a model by other universities as they've begun to introduce bioethics. People talk about Christian guidance as "trying doors." I tried this door, it opened with a vacuum behind it, and I got sucked in!

I decided that bioethics was worth giving quite a bit of time to, for two reasons. First, it is important to get people thinking about ethical issues, and if they have a faith then to help them to think properly about it. I really did see it as part of my responsibility as a Christian. The second reason was to try to show students that science is not value free, and is not done in a vacuum: it's part of the activities of society. The students loved the course, so it was also very rewarding. For quite a long time I kept these bioethics activities running alongside my research on DNA. Three years ago I gave up my lab-based research but I do supervise some students doing research projects in bioethics. We've got a large number of young people at Exeter University, so we started recording their attitudes in voluntary surveys. It has led to some fascinating and useful insights into what people think about ethical issues.

Another subject that I have often spoken about is the environment. I think people should take seriously what scientists are saying about the way the world is going. They need to realize that we're overusing many of the earth's resources. Marjorie and I have three grandchildren now, aged between five months and five years, and that makes you think about what it's going to be like for them at the other end of the twenty-first century. You hear fishermen saying, "If I stop fishing to let the cod recover it will destroy my livelihood." We need to try to get the message out: okay that's your livelihood, but what about the fact that there may be no fish at all for your grandchildren? The same is true of climate change. We may be very reluctant to change our lifestyles, but if we don't all change then our grandchildren will have no "lifestyle" at all because they'll be coping with the effects of dramatic climate change.

One of the reasons why I worked with plants is that they are the basis of the food chain. They are the prime producers, and a large proportion of the world's population relies on them totally for their food intake. It is plants that will feed the world, not bigger cattle. I don't think you can guide research specifically to technological applications all the time. If there's no basic "blue skies" research, then there are eventually no new data to apply in order to develop new technologies. So I've not done genetic manipulation specifically for crop applications but some of the work I've been involved in has led to a better understanding of how plant cells grow. You can see how some of the things we've discovered could be used to modify a plant's growth. A classic example is that if you could shorten the time to flowering in wheat by one day it would grow fifty to eighty kilometers further north in Canada, increasing the amount of land available for feeding people.

The two strands of bioethics and environmental awareness run together, because they're all about the way that we use God's gifts. Sharing of resources with the people of the world; use or overuse of the world's resources; polluting the world; using the gifts of science – they're all part of the same spectrum. There are other areas that can link in with bioethics. We need to make it very clear that the way things are done in the world at the moment is not enabling people to use its resources as God intended. We have an overall flow of wealth from poor countries to rich countries, and scientists need to say something about that. Scientists need to say something about treatment of HIV / AIDS in Africa, an area where science and ethics are very interwoven. It's a scandal that the scientifically approved drugs are not being made available there. As a Christian I want to do things right, I want to honor God and love my neighbor.

For most modern science and technology the Bible gives us no specific guidance, and you wouldn't expect it to, you've got to work it out for yourself. Christians will differ, but the key thing in these gray areas is that they recognize the validity of the other person's viewpoint. I don't think you can do anything about it, I think that's just how it is: on particular issues we will reach different conclusions. The ethics of Jesus is not about ticking boxes, and I think that our Christian influence in society should be much more about the way we treat each other. An example might be someone putting in an elevator and ramps for wheelchair users. They have ticked the boxes to show that they've provided for wheelchair users, but they may not actually care for them. Jesus would say that you should actually go and care for that person rather than just carry out a series of actions which you're required to do. I think that's really what living as a Christian means.

Science itself should also have its own set of ethical standards. Some of these are more general, such as the way I treat my colleagues, the way that I take responsibility for my actions or the way that I feel responsible to do a good job for those who are funding me. There are also the ethics of science itself: that you don't make up results, and you don't lie about your data. One of the most insulting things that was said so frequently during the height of the debate about GM crops was the accusation made time and time again by the opponents of GM that scientists were lying. That's a very big insult for a scientist, because if scientists lie then science will collapse. On the other hand, as people, scientists are no more or less ethical than journalists or than the opponents of GM crops. There have been scientists who have falsified their data and who've lied to get a job. These people get found out in the end when other labs try to build on the "knowledge" from these falsified experiments. It usually takes several labs to try reproducing the "data" before the fraud is discovered, which is both expensive and frustrating. Thankfully the ethic of truth-telling is alive and well in the vast majority of labs around the world, so scientists can trust each other's data and do make real progress.

Questioning Faith

Some of the questions that science raises are the same for everybody. What is this universe all about? If you actually stop to think about it in a deep way, the very presence of the cosmos is a puzzle. The biggest question of all is, "Why is there something rather than nothing?" That leads you into all sorts of philosophical black holes, and has to be faced. Where did the universe come from? Is there a God behind it? These are questions which everybody needs to think about, not just scientists.

There are always some people in a university environment, as in any other workplace, who are very anti-religious. They have reached a conclusion on how things are, and they don't understand how other people can disagree with them. I've found in general, though, that people are accepting of my faith in a "John's a Christian, I'm an agnostic, that girl there is a Buddhist," sort of way, and that can lead to interesting conversations. People of any age are often still discussing the big stuff of life inside themselves or maybe with people close to them. There have been a few exceptions on the way, and there is also an expectation that Christians will behave differently – that is quite obvious. There have been times when I was very aware that I was being watched. There was a colleague in Cardiff who was probably more influential than he thought. He'd certainly thrown away any idea of faith himself, and he challenged my faith a lot, particularly in terms of Christian behavior. Sometimes that's a good influence; it makes you work your faith out again.

Some scientists cannot get over the fact that you can't really prove God exists. Science is only concerned with things that can be measured, observed and experimented on: subatomic particles, galaxies and everything in between. Most scientists that I know who are not people of faith would say, "I'm not sure whether there's a God or not but my science doesn't help me to know." Sometimes that lack of definite proof leaves you feeling a bit insecure. But on the other hand you can see so much evidence for his existence.

My experience of life is that there are things that are not explicable in the way that science might describe something. There are things that appear to exist at a different level than the mechanisms of science – a spiritual level. If you believe that God is Spirit (meaning that he's not contained physically, unless he chooses to be in the form of Jesus) then I can see absolutely no problem in trying to put yourself into the presence of that spiritual being by praying. As a scientist I know someone who was healed by the power of prayer. She was quite cynical in a sense, so she wasn't the archetypal person to ask for prayer for healing but she was healed of an inoperable brain tumour. When she described the process later she said, "I don't know what it's like to be struck by lightning but I imagine it was a bit like that!" Someone could say that there was some other explanation. Cancer sufferers do sometimes spontaneously recover, that is absolutely true but not usually in such a dramatic way!

A Scientist's Life

Being an academic is quite a privilege because you're involved both in the science and with people. Everybody talks about a university not being the real world but I can tell you it *is* real. People earn their money there, going to the lab is a job, and they talk about the same things. There are plenty of opportunities to move outside of that world as well. For eight years I was a visiting professor in Molecular Biology at West Virginia State University and spent part of every summer there. I was involved, along with one or two other British colleagues, in developing courses and research that related to situations that they were facing in West Virginia: land degradation, agricultural wastes, and other environmental issues. We helped them to develop a course in "Agricultural and Environmental Biotechnology." This was one of their first postgraduate courses so it also meant that they fulfilled the requirements to apply for university status (until then it was a college). It was exciting because West Virginia is the second poorest state in the Union, and many of the students are the first in their family to attend university. Sharing that with the ethnically mixed academic staff there has been a real pleasure and a privilege.

There is also a real privilege in doing scientific research. My main interest is the replication of DNA (the way that cells copy their DNA instruction manual and pass it on to new cells). When Watson and Crick described the structure of DNA they almost thought that it might replicate itself, assembling a new DNA strand using another as a template. Now we know that mini machines called enzymes (enzymes are one type of a class of molecules called proteins) have to do it. The number of enzymes and other components that we've had to add to the list of things involved in making DNA over the years has gone up from about three to over forty. Each protein has to work in concert with the others in the right order and the right place. DNA replication is like the "failsafe" mechanism that is set up for nuclear weapons, where in order to press the button to fire them you've got to have the whole of NATO agreeing. To start replicating a DNA molecule, firstly you've got to mark the places where copying will begin, and that requires six proteins. You've then got to activate the places where it's going to start and that takes a whole series of proteins, I don't think we know how many. Only then will the DNA helix begin to open so that you can copy it, and so it goes on. It's just quite extraordinary and it's beautiful. It's more complicated than you could ever have imagined and is a confirmation of God's creative power.

Science also tells you that there are things we don't know, and maybe we'll never know, so it keeps you quite humble. I don't think, for example, that we will ever know how life started. You may have a feasible mechanism but it won't tell you how it really happened, because we can't re-run the experiment. There are gaps in our understanding of how things work; things that we haven't discovered yet. Some of these will remain gaps forever, so they should give us humility. They're not gaps in the explanation for what lies behind it all, so they should not inform our sense of who God is. If God is the Creator then he's the overarching power behind it, and covers the gaps *and* the knowledge. I think humility, awe and wonder are all rolled into one.

When you've made a new discovery, depending on how long you take to tell people in the coffee room, there's a period of time when only you and one or two other people know this. It's sitting in front of you; here are some new data which nobody else has ever seen, and it means something completely new that nobody else has ever realized. You have to add to that the experience of being a person of faith and thinking, "I'm seeing God's work here as I uncover these new facts." Being, for a short time, one of just a handful of people outside God who know that information is a real privilege. I'm not going to deny that an atheist feels awe and wonder, because they do. I think that is just increased when you realize that these intricate mechanisms you're seeing are the work of an awesome Creator.

I am always greatly encouraged by those experiences in which we encounter something above our material existence – some people call this encountering the "numinous" – I prefer to say becoming especially aware of the spiritual dimension of our existence, even being particularly conscious of the presence of God or perhaps having an enhanced sense of his glory. Another way of thinking about this is when we reach a deeper feeling of wonder that the "He who put the code in DNA" invites us to call him Father or that the eternal Word, Jesus Christ, has become our brother. Sometimes these moments of awareness are prompted by a specific event that may remind us of the wonder of the creation and the awesomeness of the Creator. Many of these experiences lie outside the realm of science. I have selected two pieces of writing to illustrate this.

> In the beginning was the Word, and the Word was with God, and the Word was God ... Through him all things were made; without him nothing was made that has been made. In him was life, and that life was the light of men. The light shines in darkness, but the darkness has not understood it ... He was in the world, and though the world was made through him, the world did not recognize him ... Yet to all who received him, to those who believed in his name, he gave the right to become children of God ... The Word became flesh and made his dwelling among us. We have seen his glory, the glory of the One and Only ... (John 1:1,3–5,10,12,14).

> ... in the mountains I felt security and joy and a oneness with nature. One day as I stood on a hill above our village ... it seemed to me that I heard a voice quite distinctly calling me by name. Looking round and seeing no-one I felt suddenly foolish and laughed rather nervously. Then I heard the voice again: "... follow me." That was all ... no thunder and lightning – just the wind rustling the heather. I lay down on my face and said quite simply "Lord, I will follow." As I said, I saw

nothing. But a feeling of awe and presence passing over me caused me to be there for some time, afraid to open my eyes.

An anonymous participant in a research project on spiritual experience carried out by the Hardy Centre in Oxford. From Rodney Holder, *Nothing but Atoms and Molecules?* (Tunbridge Wells: Monarch, 1993).

5

Life in the Lab

BILL NEWSOME

Professor of Neurobiology,
Stanford University School of Medicine

How does faith inform the work that scientists do in the lab? The way they teach their students? Does science change the way they express their faith? Does faith make a scientist more open to new ways of looking at things? How does a senior scientist in a leading university go about explaining their faith to other scientists? As we discover more about the brain, does it bring us nearer to or further from a relationship with God? Bill Newsome honestly describes how his faith led him into a fruitful career in neuroscience and the impact it has on his day-to-day working life.

I came into neuroscience because it seemed to me that this was the one area where my scientific and my human or religious interests overlapped. One of the deepest mysteries to me is the mystery of consciousness. A fundamental assumption of neuroscience is that all of our mental lives ultimately come out of the brain; if you don't have the brain you don't have mental life. Emotions somehow have a life independent of neurons (brain cells) but even these are connected in some deep and profound way to the activity of neurons in our brain. If you electrically stimulate the brain in very specific areas you can cause animals and people to experience anger, sadness, rage or fear, yet we have no good theory for why some collection of neurons should produce emotions.

I've spent a lot of my career studying the way that properties of certain parts of the brain enable us to see, and understand what we see. How in the world does the electrical activity of the eyes and brain get converted into a personal experience of being in a room? That's a profound mystery – as much of a miracle as anything that I know of in biology.

I'm fortunate to have been able to make a living at something that I am so well suited for but that is not my ultimate identity. My faith is the strongest part of who I am, what I'm about, the way I understand myself and the world. I love my science and I've been very gratified to be successful at it but if someone put a gun to my head and said, "Choose your science or your faith, Bill," there is no doubt that I would choose my faith over my science because my faith is more fundamental and a deeper part of who I am as a human being.

Life Journey

I am the son and grandson of Southern Baptist ministers, and I was raised in north Florida, where my family has been for five generations. My father was open and interested in scientific enquiry (he had a master's degree in history) and my mother was a schoolteacher. We always had a *National Geographic* around the house, and my father collected fossils, arrowheads and lots of other prehistoric artifacts. It was before the current inflamed situation between science on one hand and religion on the other, and neither of my parents ever posed any questions for me in terms of being a scientist and a Christian at the same time; that was just never an issue for me.

As a child being raised in a religious family, religion was part of my life like mother's milk almost, so certainly the fact that I'm a Christian today has something to do with my parents. My family origin does not determine my faith though, because I can show you lots of preachers' kids who rebelled against a faith that they found suffocating, threw it off and went a completely different way. I think for me, being a Christian comes in stages as life presents challenges to you. In high school, if you're bright and are at all sensitive to what's going on in the media and scholarship around you, you start asking questions and wondering, "How much of this stuff do I actually believe in, and how much was passed onto me by my parents?"

As a college student I inquired even more. I was fortunate to have some teachers who thought very deeply and coherently about religion and science, and who attempted to reconcile the two. I didn't have religion teachers who rejected science, and I didn't have science teachers who rejected religion out of hand but I had teachers who helped me explore that ground for myself, and I really appreciated that. A lot of critical thinking went on during that time. I don't think I would ever be able to put my finger on a specific moment and say, "I definitely became a Christian then:" that would be artificial. In the Baptist tradition there's a public profession of faith and baptism, and the whole idea is that you can put your finger on that and say, "That was the day." I was baptized when I was seven years old, and it was as authentic as possible for a seven year old but I think my whole life has been an unfolding act of faith. There was never an extended period of time where I stood outside the faith and considered myself a non-believer, although I think all of us have moments of doubt or non-belief. All of us struggle at times about what we really do believe and whether God is present. I think ultimately I am a Christian today because my faith makes the best sense of the world and my role in it, both at an intellectual level and a deeply personal level.

Do Scientists Doubt Differently?

I pray regularly. I don't think I could ever stop praying. When I pray for health, for a person that I love or for someone who perhaps I don't even know, I am in relationship with God. I don't believe that my prayers are going to necessarily stave off death for someone that I love. My experience of this world is that we all die and that in part suffering seems to be a mechanism that brings us wisdom and maturity. I do believe that Jesus healed and that people who were around Jesus experienced him as a healer in ways that passed their understanding.

In one sense scientists probably do doubt in a particular way because we think a lot more about mechanisms than other people. The basic mode of science is to look at something and say, "What's the underlying mechanism that makes that work?" So scientists would have a particularly hard time with exceptions to known mechanisms, as would be the case in a miracle. In that sense scientists do, I think, express their doubt or skepticism in different ways from non-scientists.

On the other hand, if you read a story in the Bible such as Jesus healing the man who was blind from birth,[1] a non-scientist might say, "Oh, Jesus just made the cataracts go away." For a neuroscientist like myself, though, I know that you have to learn to see early in childhood when the brain is developing. For people who are blind from birth and don't have surgery to remove their cataracts until they are adults, vision is nearly useless. A scientist looks at that healing and they actually see a bigger miracle than most non-scientists do, because all those connections in the brain have to be changed just right. So I'm open to the possibility that God changes the world in ways that don't go along with normal scientific mechanisms. When I pray, I pray sometimes with that outcome in mind but I don't view it as something that can be tested scientifically.

Although the writer C. S. Lewis, who is very rational and logical, was important for me during college, later in my life I found other Christian writers to be much more indicative of where I am and how I approach faith at an emotional level. At this particular juncture in my life I find C. S. Lewis less appealing than people like Frederick Buechner, Kathleen Norris and others who write from a deeper wellspring of experience and intuiting truth about life, much more than logical demonstration. I think I've left behind a point where I expect or even desire logical demonstration. I look for beliefs and a view of the world that make sense at a deeper emotional level than purely analytical logic. At the same time I don't want to be all emotional and abandon logic and rational thought. I'm really looking for some kind of mixture of the two where you can be modern and intellectual and yet be open to emotional meaning that transcends the logic to some extent or at least complements the logic.

A Life in the Lab

My job is very demanding and the further along I go the more demanding it becomes; the administrative load becomes heavier, and yet if you are going to continue to do science it has to be done to very high standards, especially in a place like Stanford. I travel quite a bit and it makes it hard for me to be a real living member of a Christian community at the depth and level I would like. When my kids were young I had an easier time with that, family was built around church to some extent and I had fewer administrative obligations. It was a less demanding period of my career and I was really concentrating on the research. At this point I'm the head of a department, I've been director of a graduate program and I'm on various committees. I can't cut back the hours in the lab, so I just work more hours, and that makes it harder for me to be a real part of the church community beyond just attending on a Sunday. I miss it, and I hope to get back to it.

Over the last ten years at Stanford I have been part of a group of Christians that have met for breakfast on Friday mornings. We read a psalm, pray and share our life journeys together. That's been a great experience for me, because before this my church life and my scientific life were basically separate. To have a group who are both my professional colleagues and Christian believers has been a great integrated activity for me, and in many ways that's been one of the most steady and meaningful experiences of church that I've had over the last ten years. There's been a very deep bonding and sharing of life with some of the people in that group. A key member of our group died of pancreatic cancer about four months ago, and we all went through that journey with him over a period of six months between diagnosis and death. Living life together at that level matters: it builds community and an understanding of what it means to be faithful to Christ in an academic community in the face of life and success, and in the face of death.

I'm also the faculty advisor for the InterVarsity[2] Christian graduate group here at Stanford. I find that a lot of Christian graduate students feel guilty at times about the privilege of working in academia. They are working really, really hard in this process of education as a scientist where you focus on smaller and smaller things. One day they wake up and realize that they are spending up to fifteen hours a day investigating this one little molecule and the role that it plays inside the cell. They say to themselves, "Is this what my life is all about? Is this really it? I'm smart, I'm talented, I have faith in God and I'm concerned about this big world we live in. Might my talents not be better spent teaching at a high school in an urban setting or being a doctor saving people's lives, rather than looking at this little tiny molecule?" This is a real issue for lots of graduate students but for some reason it was never that much of a problem for me. I think my parents taught me at a very early age that God gifted me with intellect for a reason and that learning things about the world is part of God's gift to humanity. You can praise God through scientific investigations of how his world works. As a child it was instilled in me that Christians are needed in all walks of life. I deeply inhaled the apostle Paul's metaphor of the body: the eyes don't say to the feet, "I'm what the body is all about, you're less important," and the feet don't say the same things to the eyes. I always figured I was a part of the body.

When I'm in the lab there's actually nothing about being a Christian that makes me a better scientist at that everyday level. I don't think there is anything other than my general motivation that makes me uniquely qualified as a Christian to answer a particular scientific question or do a particular experiment. I think my faith matters much more in how I run my lab and the commitments that I make to people, especially students, and the value that I place upon teaching. Sometimes my personal value commitments lead me to go the extra mile with students, where my rational deliberation as a scientist and a lab manager would say, "I should cut my losses here, I'm in a competitive situation, this student isn't producing, I've given them lots of chances, so lose the student." I tend to go another mile and work with people to try and get them to grow and understand the scientific world. I think that sometimes the pace and progress of the science in the lab has suffered for that but for me taking a student into my lab is not a utilitarian thing, it's a commitment to a person, and human relationships have a bigger value for me than the scientific progress in the end. I'm not a saint and I'm not unique, I have colleagues who are not believers who are probably just as human-centered in their values as I am but I know, for me, my Christian faith leads me to act differently with people than I would if I were not a Christian.

Does Faith Change Science?

I feel myself grappling with issues where I think my Christian instincts are aligned with good scientific instincts. Neuroscience at this point in time is consumed by breaking things down: for example you break a cell down into all its molecules, and that understanding gives you control over the cell and makes things that once appeared to be magic appear rational. This is "reductionistic science." But then there's "synthetic science," which is more like engineering. Engineers are accustomed to putting components together and seeing interesting behavior emerge in ways that you didn't predict. This is the idea of "emergence" in complex systems: you put things together and phenomena emerge at a higher level that are extremely difficult and perhaps even impossible to predict from knowledge of the lower level alone.

Emergent traits like emotions come into existence that have to be understood in and of themselves. They have relationships to the lower levels but they have an existence that is independent of them too. In the same way, the computer is more than simply the sum of a bunch of transistors and the connections between them. In one sense a computer is only the transistors and the connections between them, but describing it like this loses something in the process. You can't really understand the computer unless you understand its functional relationships, what it's trying to accomplish, and the purpose that it serves to its human creators.

My faith makes me more open to the view that these emergent properties are real and not just to be explained away. I think that's reality. Being a Christian doesn't give me a unique window onto that truth, it perhaps just predisposes me to be open to it. Neuroscientists I think need to be informed more by that kind of perspective.

I would say this to religious people, too, that an understanding at a mechanistic level does not reduce the importance of the human person. As a Christian I think that the human organism is more than simply the sum of its neurons. Our current reductionistic science is impoverished. It's not wrong, but it just loses a level of reality that we will have to recapture in the end. It's truly a product of sound reflection on the world to realize that organisms call into creation phenomena that are not usefully reduced to their component parts.

I don't follow that and say, "Therefore there's a God." I don't believe in a "God of the gaps" arguments. But I do believe that complex organisms achieve a level of existence with complex organization that is very, very difficult to explain in terms of the lower levels. This fact doesn't prove anything religiously speaking but it is consistent with the idea that a Creator made the world in a manner that would lead to the evolution of life and eventually to beings like us.

Does Science Change Faith?

Up until 1997 I was actually fairly private about faith matters within academia. The people who I worked with in my lab knew that I went to church every Sunday, so I'd have conversations with them about it but I wasn't involved in discussions on a wider level. Then in 1997 I was asked to do a talk at the Veritas Forum[3] here at Stanford. I really didn't want to. I'd never stood up in a public academic forum and talked about faith issues and I shrank from it. In some ways I think I did not want that level of complexity to enter into my relationships with my academic colleagues but in the end I decided that all of my reasons for not doing the talk were unworthy, and I decided to go ahead and do it. I probably worked harder on that talk than I have on just about any other talk in my life, trying to be real and thoughtful without being preachy or dogmatic.

In retrospect I'm really glad that I gave that first science-faith talk, because it opened up room for a lot of conversations with faculty colleagues and students in particular. A few were negative but most of the reactions were overwhelmingly good and positive. A lot of my colleagues are agnostics or atheists but not aggressively so, and they are very thoughtful, searching for some sense of purpose to the universe and trying to make sense of it. Even though they have different conclusions than I do right now, I regard them as very real seekers. I think being "out there" as a Christian in academia has been positive for me.

Some of my colleagues were militantly anti-religion and just disgusted by it, and they let me know about it. When you burrow down deep you typically find that people who give off all that emotional heat usually have some very unpleasant experience in their background that is the cause of their anger. It's interesting for me, when people talk with me honestly, to try and find out what that is and where it comes from. Sometimes I think it probably comes from bad and mistaken sources, even though they are real in those people's lives. Sometimes it comes from pretty good sources, ones that I have some sympathy with.

I find that scientists can be rigorously skeptical about certain things but then almost touchingly naive in believing that they don't make faith commitments. When I ask some of my friends in science if they are religious, they say, "No, I'm a scientist," implying that if you're a scientist you can't be religious, and that these things are opposites. They would like to believe that scientists don't take things on faith but on evidence. So I frequently ask my scientist friends, "Are you married? Did you do any kind of scientific experiment to determine whether it was a good idea to get married? Do you have any scientific basis for believing that this marriage is going to last for a lifetime? What kind of evidence did you use?" Of course they hum and ha, and eventually allow that probably all people make commitments out of deep expressions of faith, based on personal values that don't come out of science. This is sometimes a rather disconcerting experience, for a scientist to realize that she or he is a person of faith after all. I find many scientists don't actually reflect very deeply on the nature of human value, human purpose and commitment, because if they did they would not glibly put religion and science in opposite camps. Science simply does not address the most important questions in life.

I like this quote from Buechner because it combines modern scientific awareness (evolution, for example) with a very biblical sense of faith in God's saving love for each of us. Buechner draws out the basic truths of Genesis 1 about God's love for his creation (including each of us) in a way that is deeply moving to me.

Then God said, "Let us make man in our image, after our likeness, and let them have dominion over the fish of the sea, and over the birds of the air, and over the cattle, and over all the earth, and over every creeping thing that creeps upon the earth." So God created man in his own image, in the image of God he created him; male and female he created them (Gen. 1:26–27).

The passage from Genesis points to a mystery greater still. It says that we come from farther away than space and longer ago than time. It says that evolution and genetics and environment explain a lot about us but they don't explain all about us or even the most important thing about us. It says that though we live in the world, we can never be entirely at home in the world. It says in short not only that we were created by God but also that we were created in God's image and likeness. We have something of God within us the way we have something of the stars.

... I believe that what Genesis suggests is that this original self, with the print of God's thumb still upon it, is the most essential part of who we are and is buried deep in all of us as a source of wisdom and strength and healing which we can draw upon, or with our terrible freedom, not draw upon as we choose. I think among other things that all real art comes from that deepest self ... I think that our truest prayers come from there too, the often unspoken, unbidden prayers that can rise out of the lives of unbelievers as well as believers whether they recognize them as prayers or not. And I think that from there also come our best dreams and our times of gladdest playing and taking it easy and all those moments when we find ourselves being better or stronger or braver or wiser than we are.

Frederick Buechner, *Telling Secrets* (New York: HarperCollins, 1991), pp. 43–5.

6

Thinking Technology

ROSALIND PICARD

*Professor of Media Arts and Sciences, MIT; Director of the
Affective Computing Research Group at the MIT Media Lab;
Co-director, Things That Think Consortium*

*Society has been transformed by the digital revolution. What are
the best uses of this technology? Can we use it to communicate in
better ways? Can we make machines that interpret our emotions?
Rosalind Picard's story describes how curiosity led her into both
Christianity and science. Her interest in the way that people
think made her ask new questions about the way that we use
technology.*

Discovering God

I was not a person of faith early on. I was not even a doubter; I
was a proud atheist. I thought that rational scientific thinking was
the only way an intelligent person would operate and I assumed
it was sufficient for all truth. When I was twelve I decided to
lead the evolution team in the evolution versus creation debate
in a science class. To me creation was very clearly a myth and
evolution was very clearly the rational scientific view. I had
pictures of monkeys turning into people, so of course that had
to be true. My group said, "Look, this is what the encyclopaedia
says, this is science." The other group said, "The Bible teaches
it, and I believe it." They were led by the most popular girl in
the class who had blonde pigtails, a great tan and a house with
a swimming pool. Everybody liked her and I was the little geek,
so my team lost, just like that. I thought, "What's going on here?
We had the better arguments!"

At about the same time some neighbors, whose children I babysat, kept inviting me to church. I didn't want to go because, first, I wasn't interested and, second, I would have to wear a skirt. I kept faking stomach-aches and other things every weekend when they'd ask if I was ready to go. Finally my neighbors said, "It's not that important that you go to church. What's most important is that you know what you believe." They challenged me to read the Bible. I wanted to be a well-educated person, and I knew that the Bible was an important and famous book; so I decided to at least read it, figuring then I'd be better at shooting it down.

Some of my school friends at the time were reading a very modern translation of the Bible called "The Way," which had little boxes that you would tick after reading each of the chapters in the Bible. I decided to read through the whole Bible in a year, ticking off three chapters in the Old Testament and two from the New Testament each day. At my neighbor's suggestion, I started with Proverbs. I thought it would just be fantastical stories of beings appearing in the middle of air and things like that but I actually found it wise – profoundly wise.

I went on to read the whole Bible. Of course I just wanted to make sure that I knew what Christianity was about, but things started to change in me as I read it. I can describe it as an experience of being spoken to. When you enter into a conversation with somebody, if you're willing to truly listen, then you are also open to being truly changed. In this case I was gradually changed through the process of reading and "listening." It caused me to ask questions that, frankly, I hadn't really considered deeply. I had been content to dismiss religion largely based on what I had seen in the media, not realizing how superficially inaccurate that representation could be. The Bible made me think, so I kept on reading and learning, re-reading it from start to finish.

I didn't want to admit that my beliefs were changing and I still didn't want to go to church. I didn't want to be associated with religious believers, because in my mind they were just not an attractive group to be connected with, and yet I was gradually becoming one.

Having had the experience of being influenced by reading the Hebrew (Old Testament) and Christian (Old Testament plus

New Testament) Scriptures, I thought I should read Scriptures from all the other major religions too. I didn't want to be biased to believe in something simply by having been raised in the United States and Europe and reading the Bible. Of course Christianity did not start in the West; it started in the Middle East and spread, and today it is most rapidly growing in Asia but I did not know that at the time. I was able to arrange for individual study on world religions through our high school's gifted program, and I started studying all major world religions, reading their scriptures, visiting their mosques and houses of practice, and meeting with their people. I wanted to open myself to being even more persuaded by them than I had been by the Bible. I still disliked so many things about local religious culture that I hoped the other world religions would show me something better. I spent the year studying and today I continue to learn; it is important to me to always remain open and learning. However, to date I have found nothing at all like the things that I found in the Bible, with its truly culture-free message, historical evidence and ability to bring about redemption and forgiveness.

When I went off to college I met several smart people who asked me not just, "What do you believe?" but, "Are you truly committed?" I realized that there was more to Christianity than just believing Jesus was who he said he was, and that he did what people saw him do. People are not just called to believe that Jesus was God's Son, sent as a sacrifice for us, for our redemption. We are also called to follow Jesus: to try to be more Christ-like. Choosing to decide whether or not I was willing to give myself over to that commitment was another moment of major change. Jesus said, "If you are not for me, then you are against me." How could I be against him when he had done so much for me and for every human being? He was more wise, generous, self-sacrificing and divine than any other person throughout history. Is there any greater individual to follow? I wasn't looking for anyone to follow, and I was quite happy with my life, not looking for change. But I decided to run an experiment: "OK Jesus, I'm no longer going to be lord of my life. You are." The change in my life was so tremendous, and so eye-opening, that I cannot find words that are adequate to describe it.

Discovering Science

At first I found science very boring at school because we just sat there and memorized things. When I finally conducted experiments in chemistry something always blew up or went wrong. Biology specimens stank. And there was my effort to lead a team in the evolution debate that failed. Later on I discovered that we could get out of classes for a couple of afternoons a week to study at the science center across town. This was an opportunity to skip school without actually getting thrown out, so I signed up. At the science center we met a PhD student whose sense of fun was contagious. He taught us to make electrical circuits, and when I asked him how holograms were made, he lent me a laser and everything I needed to make my own! It was much more exciting than school, and my curiosity to learn about how things worked just exploded.

After high school I earned degrees in electrical engineering at the Georgia Institute of Technology, and then electrical engineering and computer science at the Massachusetts Institute of Technology (MIT). I went on to join the faculty at the MIT Media Lab, where we develop all kinds of new technologies to invent a better future: advancing human learning, expression and ability. Initially I worked on "vision and modelling" research, trying to teach computers to see and interpret the information coming in through cameras. A group of us collaborated with the world's foremost experts on human vision, trying to discover what goes on in our brains when we see. How does our visual system figure out what all these patterns of light mean? How do we think? How does all this processing fit into our skulls? Interestingly, the more we learned, the more new questions arose. Through this experience I became increasingly in awe of how the brain actually works. Our group of collaborators represented what many considered to be the top institutions in the world, with access to the most powerful resources; we were constructing the most innovative models, and yet with all the brightest ideas and best technology, we couldn't make anything that would reliably perform tasks that a small child could carry out effortlessly.

As I learned more about the brain I kept bumping into the role of emotion. That wasn't something I wanted to be associated with because I was a woman trying to prove myself in science, and who wanted to bring up the topic most closely associated with being irrational? I figured I would have been taken more seriously standing up at a scientific meeting proclaiming the merits of lip cosmetics: at least I could have described their chemistry. Like my colleagues I had thought emotion was only important for when people "get emotional," but that turned out to be only part of the truth. The latest findings in neuroscience were showing that emotion played critical roles in rational behaviors: perception, memory, language, decision-making and other vital aspects of intelligent life. Moreover, emotion was present even when a person did not appear to be emotional. As I saw this and thought about what we were trying to get machines to do in the lab, I realized that computer scientists couldn't ignore emotion: I needed to get my community to think about it even if it would make people laugh at me. I imagined my hard-earned reputation as a scientist and engineer being ruined once I went public with "work on emotion," but it turned out to be even more surprising than that.

One place where my faith has helped me with my science is that it has made me fearless. I take it literally when the Bible says "Fear only God." I'm not going to fear what all my colleagues are going to think of me. Before God all of the most intimidating professors really aren't intimidating at all. With this perspective all fear of people vanishes. As a child I was quite nervous in front of people, detested public speaking and would weasel out of any public appearance, especially the weekly show-and-tell time at school. I would hide the object my Mom made me bring so I wouldn't have to stand up in front of class and talk. I would have cowered in the presence of the Nobel Prize-winners, CEOs, rock stars, You Tube luminaries, heads of state and other people that I have the pleasure to meet regularly these days. What brought about this change in me? I came to know that when we stand before the greatness of God we are all the same size. This is true whether or not a person believes in God. With this understanding I am completely free to research whatever seems to be important, even completely new topics that involve emotion like "affective computing," that would be quite unpopular with others. People did snigger behind my back and say, "Can you believe what she's working on?" But the Media Lab embraces crazy new ideas and, while some people shook their heads, I began to pursue work in what was to become a new field.

Affective Computing and Autism

The field of affective computing includes the development of technology to detect and respond respectfully to human emotions, and to help people to better communicate, understand and self-regulate emotions. Within these goals there are abundant opportunities to build technology that helps people. One of the things that attracted me to work in this area is that there were several problems that we had been trying to address with technology, where our solutions looked as if they could be immediately helpful to people on the autism spectrum. For example, many people have difficulty reading and responding to social emotional signals in real life. We build techniques to code and decode emotional information using computers. People can use that technology to help them learn and improve their skills.

One problem is that most of us who can read emotion do so without realizing how we do it. We don't actually understand how our facial expressions suddenly mirror those of another before we even realize it is happening. In order to teach such non-verbal skills to a person who is unable to just intuit this communication, we have to figure out how it works, and this requires a lot of high-speed precision and examples that technology can help provide. A hallmark of autism is the ability to perceive and analyze detail in a focused and systematic way. The detail such an approach provides is exactly the kind of input we need to help us program the computers to be more effective at reading emotion. There's a wonderful partnership here, with the autistic person prompting us to understand social interaction systematically, and our technology that is able to help them to do that more quickly in a social setting. This is just one of many areas in which we are working with autistic people.

I'm a technology inventor but rather than simply making better inventions, I want to make the world a better place. I find it important to collaborate with and listen to artists, critics, and people who are much more human-oriented than technology-oriented. We rub against each other, sometimes making sparks but more often making a deeper sense out of things. We join in asking: "What could make the world better?" And for each idea or design, "What's the worst thing that this could be used for and how can we avoid that use?" In designing technology that actually helps people who are marginalized by society, we've got to watch out for places where they could be harmed or abused. We're trying to work with individuals who are diagnosed with autism and hear what harms they've suffered, what successes they've enjoyed and how to help them have less of the former and more of the latter.

In all aspects of my life as a Christian, I try to listen for God's guidance and calling, and be open to being led even to places I do not want to go. I've felt repeatedly nudged in directions. Sometimes I was outright forced into taking some decisions that I did not want to take at the time but I can now see that God was providing for me in ways that I didn't understand. Now I find myself in a unique position to develop technology to help an amazing group of people: people who have been marginalized and yet have a ton to offer not only to scientists but to the whole world. It feels like an extremely special opportunity that would have been difficult to craft in a deliberate fashion.

Science, Faith and Atheism

People tend to judge beliefs by the people they know who carry them. I know scientists who give science a good name and others who give their profession a rather less savory set of names. I also know Christians who fall into these two categories. My natural tendency, you could say my starting point along with my atheism, was as a person who spent time with my family and a few close friends but otherwise wasn't interested in others unless it was helpful for me in pursuing one of my various self-interests.

Then I became a Christian. I recall overhearing a person complaining about a Christian man: "That man is so miserable and wrapped up in himself. He gives Christianity a bad name." The problem with a remark like that, in addition to it being true, is that we don't know how much more miserable, how much more selfish, how much more unlikeable that man might be were he *not* now a follower of Christ. He might be entirely charming now in comparison to the alternative. For me and those who know me, it helps to keep this example in mind. My scientific community also rewards my natural tendencies to put self first: do great work in your own name, promote yourself, promote your work, maximize your impact. But because of the powerful words of the Bible to consider others better than self, to look out for others' interests and not just our own and most of all because of the example of Jesus, I am challenged to turn from this selfish nature. On those occasions when I follow these instructions something deep within me bursts free, as if into song, with full resonance that I can describe as like a violin finally being played by a true Master.

When the web first went up in 1993 my colleagues and I created web pages for our research. Our group template had a slot for "favorite book" and I inserted "Philippians." Immediately I received emails from strangers about this and set up a blog-like personal page (but this was before blogs existed) answering questions about my faith. I knew some people would see me as a nut for being open about my faith and maybe it would even mean I didn't get tenure. But I had felt God's guidance through all the trials that had brought me to where I was, and my burdens had been carried at each step, with my anxiety replaced by peace and a sense of a much greater purpose than my own. I also knew several highly intelligent people in top positions who were Christians, so I wasn't the only nut.

I've had a huge range of responses to my openness about my faith. Some people are really happy to see when a scientist admits their Christian faith publicly, as maybe it will inspire more to open up. I've also walked in on a group discussing with incredulity that the visiting colloquium speaker (me) could possibly believe in that stuff on her web page. I smiled because I used to be right where they were – thinking people of faith had to be idiots. Now that I am a Christian I know some people will assume I have lost my marbles or maybe that I believe something just because I was raised to believe it (I wasn't). But I also know that if they move beyond such superficial characterizations and ask hard questions, the ones about real meaning and purpose, that they will see more of what I see.

I think it's especially important to ask: Why does our world exist and why do we? Why do all people, even the most anti-religious, see design in the universe? Why do all of us experience purpose and meaning? Is it possible that there is a Being that transcends our world? What is the evidence for and against the God revealed through history, through Abraham and his offspring, and the events and messages (recorded in the Bible) that are reported to be for all nations, for all people? Who was the historical Jesus? What did he say and do and what does that mean for people living today?

As I've learned more, my scientific method has informed my faith because I'm very analytical, and I question things constantly. You have to be careful as a scientist, however, that you don't fall into the trap that a lot of atheists fall into. They just assume that God must be provable or disprovable by science. In fact some of them assume that the only things that are true are things science shows. Ironically what they are doing is claiming (dogmatically) that they have the only way to truth: science. But science, within itself, cannot prove the correctness of its own methods. It cannot prove its claim to be the only way to know truth. Science cannot prove most events of history, but does that mean they did not happen? To believe that God is explainable by science is to completely mischaracterize God.

The God of most atheists really is a little green man under a desk and it is easy to debunk that God. The God of the Bible, however, is the Author of all we can know, including space and time, and the entire universe in which we exist and make science. Science is no excuse to reject faith in this God. If anything, the wonderful order, organization and "knowability" of the world, illuminated constantly by new scientific discoveries, points all the more to something so much greater than science. Opening one's mind to possibilities that science cannot yet explain is also good scientific practice. The best of truth-seekers are always willing to revise what they have believed when given new kinds of data that speak of something more accurate and closer to truth. How might a Being that brought our universe into existence seek to be revealed to us? I think the answer can be found under the microscope and in the telescope and with the tools of science, but only in part. The sacrifice and example of Jesus and the message given to everyone in the Scriptures are also revelation: given not just to scientists but to all people, of all ages and levels of understanding and ability, for all cultures, for all times. God's revelation is free and not forced upon any of us. It is given for you, as long as your mind is not too narrow to let it in.

I'll never forget Dr. Gordon Hugenberger[1] talking about his meeting with one highly-educated intellectual who was discussing doing missions work for the church. The young professional suggested that his extensive education and (quite impressive) experience had made him over-qualified for simply serving others and spreading the news of Christianity. Dr. Hugenberger thought for a moment and replied, "Over-qualified? Do you want to know who was really over-qualified for this work? Jesus was over-qualified."

This passage below challenges me because Jesus, in very nature God, more than towers over the values and practices of my educational and research community, a community where many faculty, colleagues and international visitors aspire to "god-like status" via self-achievement. One MIT Sloan business school faculty responded by hanging a large sign in her office: "There is a God and you are not Him." Today's academic culture promotes many good things in seeking knowledge; however, it also nods at selfish ambition, encourages self interest and winks at conceit. The author of the letter to the Philippians, who was also highly educated and over-qualified by the standards of his day, urges us to adopt a very different way of operating, with the attitude adopted by the most over-qualified person of all time.

Philippians 2:1–11

If you have any encouragement from being united with Christ, if any comfort from his love, if any fellowship with the Spirit, if any tenderness and compassion, then make my joy complete by being like-minded, having the same love, being one in spirit and purpose. Do nothing out of selfish ambition or vain conceit, but in humility consider others better than yourselves. Each of you should look not only to your own interests, but also to the interests of others.

Your attitude should be the same as that of Christ Jesus: Who, being in very nature God, did not consider equality with God something to be grasped, but made himself nothing, taking the very nature of a servant, being made in human likeness. And being found in appearance as a man, he humbled himself and became obedient to death – even death on a cross!

Therefore God exalted him to the highest place and gave him the name that is above every name, that at the name of Jesus every knee should bow, in heaven and on earth and under the earth, and every tongue confess that Jesus Christ is Lord, to the glory of God the Father.

7

A Deeper Logic

Ard Louis

*The Rudolf Peierls Centre for Theoretical Physics,
Oxford University*

*How do we gain reliable knowledge about the world? How do we
make sense of all the different areas of knowledge available to us?
How does faith connect with a scientific way of thinking? Ard Louis
spends his working life using mathematics to understand how the
world works. His experiences in Africa influenced his faith, and his
experiences as a student gave him a passion for finding new ways
to approach academic challenges.*

An African Childhood

My first memories are of living in Bongolo, a remote village in
Gabon, Central Africa. I was born in Holland but when I was
about two, we moved to Africa and my parents became science
teachers at an isolated little secondary school in the middle of the
bush. They were avid biologists and collected all kinds of animals
in our back garden, including a chimpanzee called Bertje, who
became a family pet. The local people in our area got most of
their meat from the forest, and Bertje's mother was shot for food
when he was just six weeks old. My father traded Bertje for his
weight in sardines and we raised him with milk from a bottle. He
became like an extra brother to my sister and me, and he got up
to all kinds of crazy antics. We ran around barefoot everywhere
and ate all sorts of fruit from the jungle, so it was quite an idyllic
existence. When I was fourteen I went to boarding school in the
Ivory Coast, and a few years later I moved back to Holland to
finish high school.

My parents had become Christians not long before moving to Africa. The church we attended in Gabon was dynamic and fast growing, and I often saw dramatic changes in peoples' lives. I remember, for example, Hilaire and Yolande, who had had difficult lives. Each had several children who were being looked after by a collection of sisters, grandparents and former partners. They became Christians, met each other, got married and decided to bring their scattered families together and care for them. This change had a dramatic effect on their children, and Hilaire and Yolande ended up taking care of a large number of other people as well. If you visit them at home today you can find as many as twenty people staying with them, many of whom need some kind of help or care. Growing up surrounded by people like this had a significant impact on my personal faith.

Although my sister and I grew up surrounded by Christian influences, our parents never put pressure on us to believe certain things. They felt we should make up our own minds about faith, just like they had done at university. I had to go through a process of testing and decision as I thought about what it meant to become a Christian myself. Several experiences stand out as having marked this process. The first came when I was fifteen, at boarding school in the Ivory Coast. I had started reading the Bible seriously on my own at that point and there I read about the "gifts of the Holy Spirit." These were marked by unusual phenomena, such as praying in an unknown language or "tongues," and this intrigued me. I now believe it's not always wise to test God in this manner, but I have to admit I was at least partially attracted to these gifts as a way of testing if what I read in the Bible actually happened in real life. Around that time one of my friends who had never read about the gifts of the Holy Spirit had an odd and (to him) unsettling experience. He was praying one night and found himself speaking in a language he didn't understand. At first he had no idea what had happened but when he told us about it, I was astonished to find that his experience matched what I had been reading in the Bible.

My interest grew even more as other students began to have similar experiences. I remember one girl who had a very severe back injury. She was in traction and about to be airlifted back home to the United States. Before she left, one of my friends prayed for her to be healed. She instantly jumped up and started running around. Though I found this incredible, I did recognize that this girl's experience of prayer and healing matched exactly what I had read in the Bible.

Not long after this, I became very ill with malaria. Two of my friends at school prayed for me and within minutes of their prayer I felt completely better. Though I had already seen what appeared to have been other healings, I was sincerely shocked. I wondered if I was deceiving myself. To test if I was well, I went down to a place in my dorm where I would often jump and try to touch a ledge high up on the wall. I ran and jumped, and I touched the ledge higher than ever before. My malaria did not return. To this day I'm amazed by this story. My experiences in my boarding school years opened my mind to faith. I think they also appealed to my scientific bent: I saw the "theory" I read about in the Bible hold up in "experiments."

The healings I experienced in my high school years were obviously extraordinary. Other more mundane "tests" of the Bible were no less significant to me at the time. For example, the Bible teaches that one of the main roles of the Holy Spirit is to help people understand what it says. Right after I had my first experience with the gifts of the Holy Spirit, and didn't really know what to make of it, I went back and picked up my Bible. I had been working my way through the book of Romans and was thoroughly confused by it. But this time it was as if a light had been turned on and I understood one of its main themes: I could not earn God's favor simply by doing good things. God shows us favor through "grace;" it is a gift from him that we don't earn or deserve. Until then I had always thought that religion was mainly about doing enough of the right things to balance out whatever wrong things I did. I don't know where I got that from and I'm sure it's not what I had been taught but it's what I believed. I realized that my new understanding of grace, which is a foundational concept in the Bible, was connected with my experience of the other gifts of the Holy Spirit – and so these two experiences helped to validate each other.

When I was in my last year of high school I began to teach myself quantum mechanics (a mathematical theory that describes the physics of very small things). Quantum mechanics is an incredibly accurate and powerful theory. One of the really interesting things about it is that the concepts you can describe mathematically don't always have parallels in our day-to-day life. A famous example is the electron, the particle that flows in an electrical current. An electron sometimes acts like a wave, and sometimes acts like a particle. So is it a wave or is it a particle or is it both? This sounds bizarre, but it isn't at all once you understand quantum mechanics. I discovered that waves and particles are simply shorthand ways scientists have developed of thinking about electrons. We need the shorthand because we don't have concepts in daily life that exactly match what is happening at the quantum level. Once you understand the mathematics you can see that, at a deeper level, they are related in a very elegant and profound way.

My fledgling understanding of quantum mechanics helped me with some theological puzzles that I had found confusing. For example, how can we have free will if an all-powerful God has control of every detail of our lives? When I read the Bible sometimes it seemed that God was in control, and sometimes it seemed that people were in control. It was helpful to discover that, in the physical world, things that appeared contradictory on the surface (say an electron behaving like a wave and a particle) could be understood by a deeper and much more elegant theory like quantum mechanics. This led me to suspect that when I encountered what seemed to be profound logical or theological mysteries, a radically different way of interpreting the problem could make the whole thing come together in a very powerful way. Just as C. S. Lewis wrote about a "deeper magic" in his Narnia series, there could perhaps also be a "deeper logic."

Student Life in Holland and New York

When I moved back to Holland to finish my schooling, I found myself in a much more secular environment than I was used to. I think my experiences in Africa made it easier for me to deal with the challenges to faith that I faced in university. As I became an increasingly serious student of physics at the University of Utrecht, however, I found fewer and fewer Christians around me who were also pursuing academia. I was part of a great church but it wasn't one that engaged very fruitfully with the rest of the university or with fields like history or psychology. I began to wonder, "Must I separate my faith from my science?" That was a challenging time.

I went on to do my PhD in theoretical physics at Cornell University. There, I encountered some of the same ambivalent attitudes to learning that I had experienced among Christians in Utrecht but I was also exposed to new ways of thinking. Through reading books like Mark Noll's *The Scandal of the Evangelical Mind* and interacting with professors who were Christians and leaders in their fields, I realized that although some anti-religious views have crept into universities, there is no need for Christians to disengage from rigorous study. There was no need to feel that my faith and my science must be kept at a distance. In fact, I became convinced that such a separation would only impoverish both.

When I was younger, I had started to think that the only reason to do science was because it would give me a chance to explain my faith to people in the lab or because it might be a way for me to have entry to a country where Christians aren't normally allowed to go. I did feel that physics was what God wanted me to do but I always expected that at some point I'd have to stop and do a "real job." At Cornell I began reading books by theologians (including Dutch ones) who explained how being a Christian means that your work is important. I didn't have to see my research as simply a means to a "spiritual" end because, for God, uncovering new knowledge could be a way of worshipping him. I could serve God by being the best possible scientist that I could be.

The field of theoretical physics does not offer many career alternatives outside academia. Physicists tend to be passionate about the subject, so competition for academic posts is intense. For that reason you have to work hard if you want to make it. Putting in forty hours a week is unlikely to be enough. That's fine and I'm very happy to put in the hard work because I love science. In fact, I sometimes think it's quite amazing that I even get paid to think about such interesting things. Over time, I've seen that I have a certain talent for physics and as a Christian I feel it would be wrong not to use it. I want to make sure I do the best I can with the abilities that I have. This is part of how my faith and my science connect practically.

As a scientist and as a Christian, I find the world to be incredibly beautifully put together. I see that beauty when I go skiing in the Alps and I also see that beauty in the mathematics behind the universe. In my fourth year at university I studied the work of Paul Dirac, a physicist who tried to combine quantum mechanics (the theory of very small things) and Einstein's "special relativity" (the theory of very fast things). If you do this for an electron, as Dirac did, you find that no matter how you try, the mathematics forces you to include a new and very strange component in the equation. We now call this component "antimatter." When matter and antimatter come into contact, the matter gets annihilated into pure energy (think of Einstein's famous $E=mc^2$ equation). In 1932, a few years after Dirac made this mathematical prediction, the first antimatter was discovered in a lab. I was completely amazed to learn that mathematics could require something so spectacular from the natural world. It sounded impossible. How could mathematics make such demands? I stayed up all night trying to find some alternative way of solving this problem but was finally forced to give up. It's hard to explain to somebody who isn't trained in mathematical physics just how incredible this is!

The confirmation of Dirac's theory through the discovery of antimatter is an example of what the Nobel Prize winner Eugene Wigner called the "unreasonable effectiveness of mathematics." He said that it is "a gift we neither understand nor deserve." This "gift" is regularly reconfirmed to me through my work. In the mathematical structure of the universe I see something that looks beautifully crafted. To me, this seems to point towards a Creator or Sustainer behind the laws of the universe.

Research in Oxford and Cambridge

My faith also affects the way that I approach science. For example, as my career has progressed I've become more thoughtful about how I choose the topics that I work on. I find it useful to ask myself, "How will God judge me at the end of my career?" I suspect that God's assessment of my labor will have little to do with the number of scientific papers that I publish or the number of academic grants I get. Rather, it will be whether I served him in the best possible way and that includes my research.

Sometimes there is a lot of pressure as a scientist to measure your work by the kind of scientific journals that you publish your work in or by how much funding you get. Both of these are, of course, valuable things. However, if they dominate, they can be a distraction from doing really original research and tempt you to work on things that are easily publishable, fundable or otherwise "trendy." In the last few years I've purposely tried to take a long view and do what I think is intellectually important work, rather than what's popular at the moment. When I first decided to do that it felt risky because I didn't yet have a permanent position and I felt as if I might be lowering my short-term (if not my long-term) career chances. If you're not in a permanent position then you are in a very vulnerable place as an academic. Of course, on a longer term if you do really good work it is likely to be recognized but I was worried that I wouldn't have that chance. However, I now have a permanent post at Oxford and am happy with the research choices I have made. Trusting God with my career allowed me the freedom to make some scientific advances that I wouldn't have been able to make otherwise.

One of the effects that science can have on my faith is that, because of my work, I spend most of my time thinking in a very specific way. The methods of science are very successful but science gains its power from putting limits on what kind of questions it can answer. For example, scientists usually study things that can be tested in a repeatable way. But many things in life can't be put into those kinds of boxes. If you're not careful, taking this approach to all of life can have a negative effect on you as a human being. The idea that the scientific method is the only way to achieve reliable knowledge about my life could make me excessively rationalistic and blind to other ways of understanding the world. There are some types of information that are important to decisions I make but that simply cannot be properly judged by the scientific method alone.

Since moving to Cambridge and then Oxford, I have been able to interact with a large number of people who think about science and faith in interesting ways. At Oxford I am involved in a dinner every term that pulls in twenty-five or thirty academics for a very lively discussion on science and faith. Many of these people are colleagues of mine from physics but there are also theologians and historians involved. These discussions haven't necessarily had a direct impact on my day-to-day research but they have helped me to think more broadly about big issues. Interacting with these academics and others through the work of Christians in Science[1] and The Faraday Institute for Science and Religion[2] has increased my conviction that the discussion between faith and science is worth exploring beyond the oversimplifications that too often characterize the public debate.

An International Perspective

People sometimes ask me how my career as a physicist relates to my international experiences growing up. It's true that the physics I do doesn't have such an obvious or immediate impact on the friends I grew up with in Africa. The latest development in theoretical physics isn't usually immediately relevant to them or to most people anywhere, for that matter. But over the long term I believe that advances in science do have a profound influence on the whole world. Many of the most important technologies in use today were built on scientific knowledge that grew out of curiosity-driven research, some of it done quite some time ago. If you work in more abstract theoretical research, as I do, then you may never actually live to see the direct effects of your work on everyday life. But the potential effects of scientific advance can, in the end, have an enormous impact on the world.

History also shows us that advances in science can also impact humanity by affecting our culture and how we think. Challenging myself to think about how science and faith relate has helped me recognize all kinds of unspoken assumptions I was making about both science and faith, and lay them out on the table for analysis. I think that such an exercise is useful in other aspects of life as well. To give an example, for the past few years I have been helping my best friend to make a documentary series on African children infected or affected by HIV / AIDS. As a result I've been drawn into some of the debates about how policy-makers, for example, try to change people's behavior. These are complicated matters but it is clear to me that a lot of public health programs that come to Africa from the West have been ineffective partly because the people who develop them are often influenced by implicitly secular assumptions about the world. Because of this, these programs have often ignored the spiritual dimension and the influence that this has in many African people's lives. Sadly, I have seen this make potentially life-saving programs ineffective. I hope that the science-faith debate could influence these kinds of matters indirectly, mainly by creating space for faith-inspired thinking in the public arena.

I think Christians can also be guilty of not thinking as carefully as they should about these broader issues. Faith doesn't need to be anti-intellectual – in fact, I have found that it can enrich my academic life. I think a Christian way of looking at the world should be much more interconnected than the traditional ways that we're often taught. An economist looks at the world only in an economic way and a psychologist looks at the world only in a psychological way. There is value in that because you need to specialize to be able to achieve the technical mastery to properly use the tools available in each of these fields. But you also have to be able to take a step back and say, "This is just one way of looking at the problem. There are other, complementary ways of looking at them." A robust Christian way of looking at the world should be able to integrate all these different ways of thinking. I hope that this perspective will help me to be constantly aware of the limitations of my own discipline and how what I learn fits into a larger picture. I don't believe you have to be a Christian to be able to do that but I have found that it has helped me. Faith and science provide different lenses through which we can, hopefully, discover better and more fruitful ways of considering our world.

I love the way this beautifully structured poem begins with a description of how nature reveals the glory of God, reflects that sentiment in the middle with a discussion of how God's perfection is revealed in the law, and then ends with a more personal application to our human lives.

Psalm 19

The heavens declare the glory of God; the skies proclaim the work of his hands.

Day after day they pour forth speech; night after night they display knowledge.

There is no speech or language where their voice is not heard.

Their voice goes out into all the earth, their words to the ends of the world.

In the heavens he has pitched a tent for the sun,
which is like a bridegroom coming forth from his pavilion,
like a champion rejoicing to run his course.

It rises at one end of the heavens and makes its circuit to the other; nothing is hidden from its heat.

The law of the LORD is perfect, reviving the soul.
The statutes of the LORD are trustworthy, making wise the simple.

The precepts of the LORD are right, giving joy to the heart.
The commands of the LORD are radiant, giving light to the eyes.

The fear of the LORD is pure, enduring forever.
The ordinances of the LORD are sure and altogether righteous.

They are more precious than gold, than much pure gold; they are sweeter than honey, than honey from the comb.

By them is your servant warned; in keeping them there is
  great reward.

Who can discern his errors? Forgive my hidden faults.

Keep your servant also from willful sins; may they not
  rule over me.
Then will I be blameless, innocent of great transgression.

May the words of my mouth and the meditation of my
  heart be pleasing in your sight, O LORD, my Rock and
  my Redeemer.

8

The Faith of a Physicist

JOHN POLKINGHORNE, KBE FRS

Former Professor of Mathematical Physics, Cambridge University,
and former President of Queens' College, Cambridge, England

> *At the age of forty-eight John Polkinghorne decided that he had*
> *been in physics long enough and gave up his position as a professor*
> *at Cambridge University to become a parish priest. He has spent*
> *much of the last thirty years communicating that science and faith*
> *are not enemies but close cousins.*

From Physicist to Priest[1]

I can't really remember a time when I wasn't in some sense a
part of the worshipping believing community of the church. I
grew up in a Christian home and although my parents weren't
the sort of people who talked about religion all that much, it
was obviously important to them. We went to church together
and I sort of absorbed it through the pores. We lived in a village
in Somerset and our local vicar was a very good preacher.
I never had any real difficulty from quite an early age sitting
and listening to him. In that sense I've never stood outside
the Christian community. Obviously as you grow up, you
begin to ask questions that you wouldn't have thought of as a
child but I've never in the course of all that felt I faced a crisis
situation – an "either-or" situation, go with science or go with
religion. I've always wanted to hold the two together.

I was a clever little boy and I was particularly good at
math. Towards the end of high school, I was taught mathematics very well and I fell in love with it. If you can do it, it's a
very exciting sort of subject. So when I went to Trinity College

at Cambridge University I studied mathematics. When I arrived as an undergraduate in 1949 I went to the CICCU (Cambridge InterCollegiate Christian Union) Freshers' sermon in Holy Trinity Church. The preacher was a chap called L. F. E. Wilkinson, and he preached about Jesus and Zacchaeus.[2] He pointed out that Jesus was on the way to Jerusalem and if Zacchaeus hadn't taken this opportunity, he would never have encountered Jesus, and he said, "You seize your opportunity." I'd been in the army before I came to university and I had kept my communicant life going but I felt there was a need for a bit of reorientation in my life. At the end he said if you want to commit yourself to Christ, come forward. I went forward with that. There would have been a time when I would have spoken of that as my actual Christian conversion. Now as I reflect on my life I see it as an important step of commitment to Christ but I think that my conversion was a continuous process rather than anything sudden.

In the course of my studies I got interested in the way you can use math to understand the physical world, so I did my PhD in theoretical physics (particle physics[3]), under Abdus Salam, a Pakistani Nobel Prize-winning physicist. That was the start of a fairly conventional career. I did some research at CalTech (California Institute of Technology) with Murray Gell-Mann who was one of the great figures in the subject, and I had a teaching job at Edinburgh University. I then came back to Cambridge, and was a university teacher for a long time. I was a professor for eleven years and saw the last of my scientific career here. All that I enjoyed very much. I saw it all as being a Christian vocation and I didn't have any anxieties about that. I felt I'd been given a certain degree of talent and it was up to me to use that talent.

The trouble is that in these mathematically-based subjects you don't get better as you get older, and I could see that this wasn't going to go on forever. You probably don't do your best work before you're twenty-five but you almost certainly have done it before you're forty-five. The subject was also changing, and becoming speculative – "string theory" and all that sort of thing. I didn't care for that so much and I wasn't so mathematically equipped to deal with it. I came to the conclusion that after I had passed my forty-fifth birthday, I'd done my bit for physics. I began to think what I might do. Of course I talked to my wife about it, it was going to be a joint decision. Quite quickly and without any real drama the idea came that to seek ordination to become a minister of word and sacrament was a worthwhile thing to pursue. In 1979 I resigned from the university and went to Westcott House in Cambridge to train for the ministry.

I had a very interesting period of about eighteen months when it was known that I was leaving the academic world but hadn't left yet, because I had to wind up my academic affairs. It's one thing to go to church on Sundays, but to give up a professorship and train for something else – that was a bit more than a gesture. It's a slightly odd thing to do. I think a lot of people realized I was a religious person but they didn't expect me to take it quite that seriously. My immediate colleagues, certainly here in Cambridge and indeed the UK, would all have known that I was a religious person but internationally it would not have been so widely known. Particle physics is an international village, and we all knew each other. People were saying to me, "Oh John, what are you up to?" They mostly weren't thinking so much about my becoming a clergyman but just the fact that I was a Christian. We would have a conversation over a cup of coffee in some laboratory canteen and I would try to explain why I had a Christian belief. Of course I couldn't say all that much in such a short time and in my mind there formed the idea of what I would have said if I'd had a few hours in which to say it.

I wrote my first book on science and religion, which is rather grandiosely titled *The Way the World Is*, during a funny year I had. I had just been ordained but I was still in Cambridge. I hadn't gone into full-time parish life because my wife was also having a middle-age adventure and was training to be a nurse. She needed another year; you have to be more careful with nurses, they train for longer ... I was a deacon[4] in the parish where we lived, and I did some part-time teaching at Trinity College. I had a certain amount of time on my hands and I wrote that little book. I loved writing and one thing leads to another, so since then I've been a fairly prolific author.[5]

Science and Theology

I'm basically a scientist with a strong interest in theology. You can't become a real professional in two spheres, or at least most of us can't, so I'm conscious of not being a professional theologian in the sense of having had that long apprenticeship and training. I like to describe myself as theologically being a "bottom-up thinker." The interesting thing for a scientist, I think, is to ask, "What are the phenomena, what are we trying to understand?" We don't know beforehand how we should think about them, so we will allow experience and circumstances to shape the way we think about them. This is "bottom-up thinking," moving from experience to understanding, and that's how I approach theology as well. For example, since I went to Westcott I've been very interested in biblical theology. I read as much on New Testament theology as I can, because here are, if you like, the fundamental phenomena that the Christian faith is concerned with. There is a sort of kinship with science there. Of course there are differences; you don't have the experimental method to hand in theology, and you can't repeat things in that sort of way but they are not as different as people think. That's one of my fundamental messages: that I'm not a vegetarian butcher. There is a cousinly relationship between thinking about theology and thinking about science.

A very striking phenomenon to me has been the fantastic sales success of Richard Dawkins' *The God Delusion*. Pretty much all the serious reviews I've read of it have said it's a bad book (as it is!), nevertheless it's sold over one and a half million copies. So there are a lot of people out there who just want to be told, "Religion is all nonsense, don't bother with it." That's a depressing thought and there's absolutely no question that the opponents of religion sell many more copies than those of us seeking to defend religion. Somehow or other we're not getting it across. I think quite a lot of people are troubled at the possible threat of religion and want to have an excuse not to take it seriously.

The Christian worldview is a fantastically profound and deeply explanatory understanding of the world. That's one of the things I always try to get across. The other thing I'm trying to get across to my non-believing friends in the academic world is that there are reasons for religious belief, and there are enlightenments and insights that come from religious beliefs that are worth considering and taking seriously. So many people have this picture that science is fact and religion is opinion; that there are reasons for scientific belief but religious belief is just shutting your eyes and gritting your teeth and believing impossible things because some infallible source of authority tells you that. I don't think that's what religious belief is like at all but a lot of people think that and they're afraid of it for obvious reasons. They don't want to commit intellectual suicide. So I'm always trying to say, "Look, you may or may not think my motivations for my Christian beliefs are adequate but they are at least there." They're something to think about. It isn't just a question of blind faith. People sometimes say to me they've read my books and found them helpful, and sometimes people say that they brought them back to Christian faith and that's something I feel tremendously grateful for. No world-famous scientist has said that to me but a number of people have said it and that's the thing that counts.

When I decided to enter the church I thought at the time that I was going off to live my life as a parish priest. I did that for about five years and there were various forms of fulfillment in it, for which I'm grateful. I hadn't really thought until I was about five years into the ministry that writing and thinking and speaking about science and religion was to be a serious part of my vocation. I was curate in a large working-class parish in Bristol and then vicar of a largish village in Kent. Neither of these were societies that were particularly academically oriented and of course there were bits of me that weren't exercised. Then I got an unexpected invitation to return to Cambridge University to be the Dean of Chapel in Trinity Hall, which was a job that had both a pastoral and an academic side to it, and it seemed an ideal combination. After a bit of hesitation about leaving the parish so soon, we did come back to Cambridge. I've been able to keep a priestly ministry in the academic world. In Trinity Hall I did of course, and later when I was president of Queens' College I played a big role in the chapel life in the college. I now lend a hand in the parish of the Good Shepherd in which I live. So that's how in some sense my professional and my Christian life have intertwined with each other in ways that I'm grateful for and have found satisfying.

One very surprising development was when somebody rang me up completely out of the blue and said, "We want you to be chairman of this committee about use of fetal material." I think I have a certain gift for chairing things. My natural desire is to seek consensus about something – not at any price – but in order to try and see if you can bridge things together. Of course the other people on the committee really were experts so I was, in a way, the "ringmaster." We were actually a very successful committee; we produced a good code of practice that is still in force. Then of course one thing leads to another and I was asked to be involved in other governmental committees. In some sense it makes it easier to "hold the ring" if you're not professionally committed to one side or another, and my experience has been good. I think that the people who constitute these committees and commissions are carefully chosen, and you don't get gung-ho scientists who say, "You know, gosh, this is the solution to everything, we must press ahead ..." People are genuinely concerned to make ethically responsible decisions. I enjoyed that period. It was a very unexpected aspect of life but my life has had several unexpected aspects.

Christians and Science

I think it's extremely important that the Christian community take on board the fact that the God who is the Creator of nature acts as much through natural processes as anything else. If you could demonstrate to me that the whole development of life on earth from the first bacteria up to us today is, in its process, entirely explicable in scientific terms, that wouldn't for a moment say to me that this was not God's creative purposes at work. There's a theological fallacy that God only works in the gaps; in the things we don't understand scientifically, which is an extraordinary thing to think really. The "anthropic principle" (the "fine-tuning" of the physical laws that allows life to exist in the universe) is about the laws of nature and that's brute fact as far as science is concerned,[6] a gap in what it can explain. The fact that these laws had to be so specific is very striking. The only way out of seeing this as the gift of the Creator to creation seems to be the rather extravagant assumption of multiple universes.[7]

There's nothing wrong with intrinsic gaps like that, where a description at one level genuinely doesn't explain the whole thing. The gaps that are misleading are gaps of ignorance, when you say, "Science will never be able to explain this." I think there's no theological mileage in – or requirement for – the idea that science at this level can't tell a continuous story of development. But if it's true that (as I believe) the "bottom-up" causes don't account for the whole structure of the universe, then we are "people of the gaps." Our powers of freewill arise from that,[8] and there's nothing improper in that.

One has to keep on saying all the time that creation theology isn't about how things began; it's why things exist. Scientific discoveries do undoubtedly modify the tone of theological discussion. There is an interaction between theology and general culture and that includes science but it's a question of the tone of discourse, I think, rather than the actual theology itself. If we are seeking to serve the God of truth then we should really welcome truth from whatever source it comes. We shouldn't fear the truth. Some of it will come from science, obviously, but by no means all of it. It will sometimes be perplexing, how this bit of truth relates to that bit of truth; we know that within science itself often enough and we find it outside of science as well. The crucial thing is to be honest.

God and Physics

It probably is true that physicists are a lot more open to questions of faith than others. I think there are a variety of reasons for that. One is that physicists look at reality at a certain level, where the world is very beautifully ordered and there is a type of natural cosmic religiosity that comes out of that. It's not unreasonable to ask if there might be some "Mind" that lies behind it. Einstein said that when he made a big discovery he felt "like a child in the presence of the elders." There is that sort of feeling to it and I think we should take that seriously.

I even think that scientists in general are more open to faith. A lot of our humanities friends think there's only opinion around and, "You have your opinion, I respect you for that, I have mine," and so on. I think that scientists do have this feeling that there's a truth to be found and that's very important to the exploration of religious belief. These things are worth exploring, yet there's a sense in which they're dangerous to explore because religious belief is much more demanding than scientific belief. I can believe in the components that make up an atom very firmly and it doesn't really affect my life that much but I can't believe in Jesus Christ without it having all sorts of effects on my life, and that is one of the problems. I remember years ago a very nice biochemist friend of mine saying to me, "I just don't want there to be a God."

People say they want "proof" for God but I think proof is a misleading category. Proof (meaning by that absolute certainty) is I think probably not accessible to human beings in any interesting area. There is an inverse relationship between certainty and things that are significant; I think you can have one or the other but you can't have both, and I think that's true of science as well as anything else. People say that science is concerned with facts and certainty and that sort of thing but of course it isn't really. Scientific activity is theory and experiment intertwined, so you need the theory to interpret the experiment. The experiment will confirm or disconfirm the theory. It's a much more subtle thing. Michael Polanyi, the philosopher of science that I'm most interested in, says in his book *Personal Knowledge* that he is seeking to understand how, "I may commit myself to what I believe to be true, knowing that it might be false," and he was talking about scientific belief. I think that's the human situation. What we're looking for is best explanation; rationally motivated belief to which it's reasonable to commit oneself, but to require absolute certainty is I think just beyond us. I think proof is a word that we should be careful about. It's rationally motivated belief that we search for, in science and religion and everything between those two.

I think you explore the final limits of science by trying to explore them as far as you can go. What you have to avoid is this over-triumphalistic interpretation of what's going on: for example, the claim that understanding human consciousness is the last frontier and the heroic armies of science are just about to cross it. That seems to me blatantly untrue, it just isn't like that. As another example, I'm sympathetic to the search for a "grand unified theory."[9] Whether we can find it or not I don't know. It's been a disappointing search in the last thirty years but it's possible. Theologically I think we might rather expect it. If there's one Creator there might be one grand plan for creation. It's when people stop talking about "grand unified theories" and start talking about "theories of everything" that arrogance comes in. Even in physics it's perfectly clear that if we understand the sub-atomic world totally we still don't understand how the everyday world emerges out of it. I do think that the "grand unified theory" is not a theory of everything and it certainly isn't a theory of you and me.

My faith has neither helped nor hindered my physics, in my view. If I compare myself to my colleagues in physics, I can't detect any difference between believers and non-believers in the way in which we do physics. Physics is likely to be the least influenced because it's the most impersonal type of science. (I guess that you might get a rather different answer if you spoke to somebody in the human sciences, such as psychology, for example.) It's only when you get beyond science and start incorporating the ideas of science into a wider and deeper worldview that big differences immediately appear. At the physicist's level it does seem to me that any worldview is going to have an unexplained starting point. Nothing comes from nothing and a materialistic worldview has to take essentially the laws of nature as its unexplained starting point – it's brute fact. It does seem to me, though, that if you look at these laws, their rational beauty, their order, their fruitfulness, their "fine-tuning," they do seem to point beyond themselves. It doesn't seem to me intellectually satisfying to say, "That's just the way it happens to be." The alternative position, that the will of a divine Agent lies behind the structure of the universe, does seem to me to be a better explanation of the richness of the world.

Colossians 1:15–20

He is the image of the invisible God, the first-born over all creation. For by him all things were created: things in heaven and on earth, visible and invisible, whether thrones or powers or rulers or authorities; all things were created by him and for him. He is before all things, and in him all things hold together. And he is the head of the body, the church; he is the beginning and the first-born from among the dead, so that in everything he might have the supremacy. For God was pleased to have all his fullness dwell in him, and through him to reconcile to himself all things, whether things on earth or things in heaven, by making peace through his blood, shed on the cross.

I value this passage because it speaks of the cosmic significance of Christ, the One in whom all things hold together and who redeems all *things* (notice, not just all people) by the blood of his cross. Here meet my deepest religious beliefs and my strongest scientific concerns.

9

Heart and Mind: Understanding Science and Faith

DEBORAH B. HAARSMA

Associate Professor in Physics and Astronomy,
Calvin College, Grand Rapids, Michigan

Christian researchers often say that scientific discoveries uncover more of God's creative power. But how do people of different faiths work together in science? How can they reach reliable conclusions? How can a scientist pray about their work? In the United States science is taught, and research is carried out, in both secular universities and Christian colleges. Deborah Haarsma has lived and worked in both of these environments. Alongside her research in astrophysics, she now spends much of her time helping students and others to relate their faith to their studies in science.

Star Gazing

It's almost a universal human experience to look up at the stars in the dark night sky and say, "Wow, that's amazing! That's beautiful!" But as a Christian, I have a deeper experience. I not only see the wonder and the vastness of the universe but also a connection back to the Creator who made it. I believe that the God of the Bible created all that we see in the whole universe and part of what God has commanded us to do is to study and take care of it.

When I was growing up, I always enjoyed math and science. In high school I began to consider science as a career. At first I thought chemistry, but chemistry labs were so smelly and messy! Biology didn't have nearly enough math in it but math itself didn't have enough science in it. What I really liked was using math to study the physical universe.

As an undergraduate I majored in physics and music. Then I went on to do a PhD in astrophysics at the Massachusetts Institute of Technology (MIT) in Boston. Astrophysics is a field where we apply the physics learned in the laboratory on Earth to the extreme conditions out in the universe. By studying those extreme conditions, we learn things about physics that are beyond what can be studied in the lab. Out there in the universe you get so many extremes, like very intense gravity near neutron stars or a vacuum between stars that has less than one atom per cubic centimeter or jets of particles that are thousands of light years long and moving near the speed of light!

I'm interested in objects that are much further away than our solar system and the Milky Way. I study the structure and origin of the universe by looking at galaxies that are billions of light years away. The universe behaves like a time machine, because light takes time to travel. When you look at something that is far away you are seeing it not as it is now but as it was when the light left it. If something is a hundred million light years away, it means that the light has taken a hundred million years to travel here. So you are seeing what an object looked like a hundred million years ago.

In my current project I'm studying galaxy clusters, which are the largest objects in the universe that are held together by gravity. These are huge systems which are typically 100,000,000,000,000 times the mass of the sun, and of course the sun has over 300,000 times more mass than the earth. What still excites me the most is when I make observations in astronomy and what I see exactly matches what I predicted mathematically. I'm actually able to write down an equation that describes the real world. I see evidence of God's glory there too, not just in the beauty of the night sky.

I see the fact that we can do science at all as a tremendous testimony of God's faithfulness: the laws of physics are stable in every time and place where we test them. There's a great passage in the Bible (Jer. 33:25), where God says he has "established my covenant with day and night and the fixed laws of heaven and Earth ..." God is speaking to the nation of Israel and points to the fixed laws of nature as a testimony of how faithful he's going to be to his people. The only reason that physics works at all is that experiments are repeatable, and the only reason they're repeatable is because God is faithful in governing the universe.

Thinking Faith

I grew up in a Christian family. The church we attended was one that nurtured and challenged my faith. It was assumed that the earth was created in six days about ten thousand years ago but it wasn't taught as "You must believe this to be a Christian." As a teenager and at college I made conscious decisions to make the Christian faith my own. During this time I wasn't asking, "How can I be a scientist and a Christian?" because I hadn't yet decided on science as a career. I was asking, "Why do I believe?" and, "What career should I choose to serve God best?"

I studied at a Christian college and the whole faculty there was an example of how you could be an intelligent Christian. They showed me how I could be a Christian with all of my mind as well as all my heart. John Polkinghorne's phrase sums it up: "You don't have to commit intellectual suicide to be Christian." There was a chapel speaker who talked about the importance of having Christians at universities. He said that we need Christians in all walks of life and being a scientist at a university is really important. I thought, "There's something I can do!" I knew it was important to share my Christian faith with other people and I had wondered if I should be a missionary but that made me uncomfortable rather than excited. Here this chapel speaker gave me another option, a context that I was uniquely gifted for, to actually go into academia and live out my faith there. Also, if you look at the population at large, not very many people can do astrophysics. To make the best use of the talents God has given me it made more sense for me to do the science that I do. Around the same time, I also realized that God needs scientists in the church too, to help the church understand science, so I see that as the other part of my calling.

When I was working on my PhD, I became interested in astrophysics (at college I studied plasma physics and didn't take astronomy). If I was going to be an astronomer, I realized I really needed to dig into the issue of the age of the Earth and figure out what I thought about it. At MIT there were students and leaders at

an InterVarsity group (a student Christian movement) who gave
me an opportunity to talk about science and Christianity issues
at a deeper level. They recommended some books written by
people with different points of view. I had to do a lot of thinking
and that was definitely a gradual process over a few years. But I
didn't feel it as a crisis of faith, it was just an issue that I had to
work out in the context of my beliefs. At the InterVarsity group
I learned from my friends and with them, and that was the first
place where I gave a talk on science and faith.

After all my reading I realized that the Genesis accounts are
teaching us important things (there is one God, God created the
world, God declared it good) but that the time-scale was not the
essential part of the story. I also realized that God wants us to look
out at the physical universe, and that's another way he reveals
himself to us. I came to the view that the Bible is teaching us the
"who" and the "why" of creation but the universe is teaching us
about the "how" and the "when."

The things that really make me doubt and challenge my faith
are not in the area of science. The questions that have bothered
me most are things like, "What's the fate of people who've never
heard about God?" or, "Why does such suffering happen and
God doesn't stop it?" I've worked through some answers to those
questions but they still come back to me now and then.

I married my husband Loren when he finished his PhD and I
was in the middle of my own studies at MIT. A few years later
we were both invited to apply for faculty positions at Calvin
College in Michigan. Many colleges in the US, including some
of our major universities, were founded as religious institutions.
Over time many of those universities have left some or all of their
religious roots behind. Calvin College still has that connection.
All of the faculty are required to be Christians, to sign a fairly
detailed statement of belief, and to be members of a local church.
Students are not required to sign a statement of belief but most of
them are there because they want their studies to impact on their
faith. Calvin College has a strong science program, and allows us
to continue our research. What it also provides is the chance to
dig into science and faith issues more deeply than someone who
teaches at a university. I'm not only allowed but I'm required to
show how science connects to the Christian faith in every course

I teach. Hopefully we can provide support for scientists who don't have that luxury.

Prayer

I have faith in a personal God, so I believe that when I pray God hears my prayer and is concerned. As he chooses, he answers. I've had prayers answered with a "No" and I've had prayers answered with an abundant "Yes." In those "Yes" answers to prayer I really do see God's hand at work, because the answer has been so far beyond what I thought would have happened by my own efforts or by ordinary circumstances. Those answers have not been violations of physical law; there's been no inconsistency with the "fixed laws of heaven" in those answers to prayer. I do believe that God can and will occasionally answer prayer with something that does go beyond the laws of nature, for example when somebody is healed from a disease. I believe God can choose to step out of his regular pattern and do something different at times, but it would be for a reason relating to an answer to prayer or something about God's desire to interact spiritually with his people. So I see miracles of healing, miracles in human history or miracles in the Bible that God used to establish his chosen people and develop a relationship with them. It seems less likely that God would do miracles in natural history that we couldn't discover until we had modern science.

I teach at a Christian college, so it makes sense that if I have a summer research student, I should pray with them. If I knew that they were skeptical of Christianity I wouldn't force them to pray but most of my students I know are Christians and have already expressed their faith to me. It makes sense for us to pray together about our work and that is a real privilege. When I pray about my science, I pray the prayer that the seventeenth-century astronomer Johannes Kepler prayed about his work. Kepler prayed something like, "If I'm on the right track, if what my fellow astronomer Copernicus said about the solar system is right, that the sun is at the center and the planets go around it, then Lord please bless my work and show me what the truth is." Every time I have a new research student I pray that prayer as we start off the summer's work: "If we are finding something true about the universe, then

Lord bless our efforts, give us the creativity and the insight to see what you have made." I have never prayed for particular results in an experiment or observation. I believe that I'm looking at the universe to see what God is telling me, and I'm not going to tell God it has to be a certain way! There are certainly times when my scientific hypothesis is that it's going to be one way and then the data shows that it isn't. All scientists have to have humility. A lot of the time your hypotheses are wrong and you have to revise them; that's the whole nature of science. If it was always the way I expected then it wouldn't be any fun, it would just be me inventing things!

The Purposeful Universe

I can understand how some scientists look at the natural world and decide not to believe in God, because I don't believe that nature gives us *proof* of God. There are a few scientists that have come to faith because science pointed them that way but I think for most of us it was something in our personal lives. It was a person we met, it was seeing Christians in action, it was finding that Christianity explains human nature, it was sensing the presence of God. Those non-scientific arguments and experiences are more important in making a religious commitment than logic and scientific evidence.

It's unfortunate that both the Bible and science have been hijacked at different times. There are people who will take scientific evidence and say, "We have a scientific explanation for X, and therefore God is not involved in X and we don't need God." There are other people who say, "Science can't explain X, so God must be involved and that's proof of God." The two sides argue against each other, driving each other to more extreme statements. I think both sides are completely wrong. Both make the assumption that a scientific explanation is opposed to God's involvement. In the Bible, though, God is described as in charge of every part of the physical world in regular patterns. In my view, science is merely describing how God works in nature. God is also present in the X that science doesn't understand yet but it's foolish to use that as proof of God's existence. God wouldn't disappear as soon as a scientific explanation for X is found.

Many people say, "The universe couldn't have come about by chance, it must have been made by God." When they say that, they're thinking of chance and randomness as processes that are without meaning, intention or purpose. But there is another, very different, meaning of the word "random." Many physical processes are random in the sense of being unpredictable, such as the rolling of dice or the motion of an electron in an atom. It's impossible to predict the outcome of these events in a scientific way. We see scientifically unpredictable processes like these throughout the earth and the universe, and some of them, like star formation or genetic mutation, were involved in the origin and development of the universe. The problem comes when you confuse the two definitions and say, "Well, this process is unpredictable, so therefore it is purposeless and meaningless." Randomness can actually be purposeful. Artists and engineers sometimes use scientific randomness in a purposeful way, to create a variety of paintings[2] or to try out system designs. Similarly, God could choose to use randomness in a purposeful way to bring about the earth and life. To say that God isn't involved in a scientifically unpredictable or random process is not a valid argument. Proverbs 16:33 says, "The lot is cast into the lap, but its every decision is from the Lord." Casting lots was the ancient equivalent of rolling dice and God was governing those events.

There are several interesting lines of evidence that our universe appears to be set up for life. The physical laws and fundamental particles appear to be "fine-tuned" so that life could develop. For example, if gravity was much stronger than it is, the universe would have immediately collapsed and stars and planets would never have formed. The fine-tuning is not definite proof that God designed the universe but it is evidence that fits with faith in a God (see chapter 8). One argument against fine-tuning is called "the weak anthropic principle." It says that we should not be surprised that the universe is well suited for life, because if the universe wasn't that way then we wouldn't be here to even wonder about it! That's certainly true but it doesn't really explain the whole story. The philosopher John Leslie suggested this analogy. A criminal is about to be executed and he stands before a firing squad. The firing squad has a dozen members who are all experienced marksmen. They all have their guns pointed right at his head, the commander says "Fire," they all fire. But the accused walks away free! For some reason he's not killed. Is he going to say, "Well of course I wasn't killed, if I was killed I wouldn't be here to wonder about it"? No, he's going to want to know *why*. What happened? Were there no bullets? What was going on? You still need an explanation. The "weak anthropic principle" doesn't really explain why and how the universe is the way it is.

One other argument against fine-tuning comes from "string theory." String theory is an attempt to connect quantum mechanics (which explains small things, like the interactions inside atoms) and general relativity (which explains powerful gravitational fields). Currently, quantum mechanics and general relativity give conflicting answers to questions about things as small as atoms that have intense gravitational fields, including the very first moments of the Big Bang. String theory suggests that there is some sort of mother universe from which many universes could form in many Big Bangs. It's a mathematical theory that hasn't been tested experimentally but it's interesting to think about how these cutting-edge ideas might interact with my faith. If there really is a mother universe or "multiverse," that would point to the existence of many universes besides ours. These universes might have a range of different physical laws. Perhaps ours is the only one where all the forces are balanced just right for life and there are many other universes out there that are sterile. If that's true, it would explain some of the fine-tuning that we see in our universe.

I could reconcile the idea of the multiverse with my understanding of God. For one thing, I'm ready to find out whatever the universe is going to show us about its formation, and I can still see God as the Maker of it all. Also, you still need the mother universe to be fine-tuned because it has to be able to produce universes with a range of properties. The multiverse doesn't entirely eliminate the fine-tuning argument. But I'm also fine with the idea that God made only one universe, even if we can never figure out what God did in the first instant. There are a few experiments that are going to come online in the next few years that might have some input into these different theories. I'm waiting to see the results of this research and what the universe itself tells us about how God made it.

Scientists and their Beliefs

Everybody brings their own perspective to science, so it's impossible to have pure scientific results without any other influence. I think most scientists today see science that way: as a human activity. We talk about "The standard model of particle physics" or "The big bang theory" or "The theory of general relativity." We say "model" and "theory" instead of "law" because we know that we don't have it all figured out yet. So how do we know that the results of science are reliable at all?

Science functions as a community, with people from different philosophical or religious backgrounds working together. In my research I've collaborated with people of many different belief systems. We can work together because we all recognize the worth of this kind of investigation. All scientists share certain philosophical beliefs: that the universe is regular in its operation; that there are regular laws; that humans are capable of understanding and describing those laws; that we need to do experiments to test those laws and make sure our understanding is correct. We might all come to those beliefs from a different angle. I come to them because I believe in the Christian God who faithfully governs the universe. An atheist might choose to believe them just because they seem to work. We can overlap in those beliefs even while disagreeing about other things, and where we overlap we're all doing science on basically the same playing field.

If many scientists with different viewpoints can get together and come to the same conclusion, that gives you some sense that they are probably on the right track. It is unlikely that they are all biased in the same way. The scientific process, including the way that scientific papers are reviewed, helps us overcome biases and get at something closer to the truth. There's a similar process in biblical scholarship, where different scholars share their work or work together. Christians read the Bible in different ways and see different meanings in the same passage, so not everybody is right about their own interpretation. The practice of correcting and challenging each other helps them come to a better understanding of Scripture.

I believe that God would not say something through the Bible that contradicts what he reveals though the created world. The conflict comes at the level of human interpretation. We could be wrong in how we interpret the Bible or wrong in our scientific interpretation of nature (or both!). As Christians, we can't deal with conflicts by throwing out science or ignoring parts of the Bible because we believe God speaks through both. Instead, our role is to continue to investigate both aspects of God's revelation and keep looking for that underlying truth.

A Christian's primary response to God's world should not be debate but an overflowing of praise and worship of the Creator. We see this in Psalm 29, where the biblical author looks at God's world and responds with praise.

Ascribe to the LORD, O mighty ones, ascribe to the LORD glory and strength.

Ascribe to the LORD the glory due his name; worship the LORD in the splendor of his holiness.

The voice of the LORD is over the waters; the God of glory thunders, the LORD thunders over the mighty waters.

The voice of the LORD is powerful; the voice of the LORD is majestic.

The voice of the LORD breaks the cedars; the LORD breaks in pieces the cedars of Lebanon.

He makes Lebanon skip like a calf, Sirion like a young wild ox.

The voice of the LORD strikes with flashes of lightning.

The voice of the LORD shakes the desert; the LORD shakes the Desert of Kadesh.

The voice of the LORD twists the oaks and strips the forests bare. And in his temple all cry, "Glory!"

The LORD sits enthroned over the flood; the LORD is enthroned as King forever.

The LORD gives strength to his people; the LORD blesses his people with peace.

Nothing in the psalm talks about how a thunderstorm works, there is nothing about meteorology, precipitation, cold fronts or electrical discharges in the atmosphere. But it's everything about "who." The storm is under the Lord's control. Every aspect proclaims his authority and glory. The poet artfully uses the human experience of being caught outside in a storm – the gust of wind, the peal of thunder, the flash of lightning – to declare God's glory and power.

The poem is also a lesson for God's people against the idol Baal, a Canaanite god believed to be the divine power in thunderstorms. Not so, the psalmist proudly proclaims; Israel's God, the Lord, is in control.

Deborah B. Haarsma and Loren D. Haarsma, *Origins: A Reformed Look at Creation, Design and Evolution* (Faith Alive Christian Resources, 2007), p. 26.

10

The God Solution?

ALISTER MCGRATH

*Professor of Historical Theology, Harris Manchester College,
Oxford University*

*How do Christians respond to criticisms by influential atheist-
scientists like Richard Dawkins who say that science has ruled out
the existence of God once and for all? Does the God that Christians
believe created the world and rules over the whole process, past
and present, slip into the background and gradually fade away as
science progresses? As scientists discover how animals grow and
develop, uncover the machinery behind the thousands of complex
functions in every cell of our bodies and learn how the stars were
formed, is there no place left for God? Alister McGrath's journey
from atheism to faith and his work in both science and theology led
him to believe that this is definitely not the case. He believes that
scientific discoveries can uncover more about our Creator than we
had ever imagined.*

Beginnings

When I was young, I developed a passionate interest in the
sciences. This began with a fascination with the night sky. I
read all I could on astronomy and, when I was about nine, I
built myself a little telescope. At about the same time, I inherited
an old microscope from my great-uncle who was head of
pathology at the Royal Victoria Hospital in Belfast. With it
I discovered the tiny creatures that live in pond water, the
incredible complexity of the parts of garden flowers and so on.
I can still remember the spine-tingling feeling of wonder when
I saw these things for the first time! I was looking at a different

world and it was like opening Pandora's box. Part of what fascinated me about science was finding out how things work but it was also about finding new things that I needed to explain. It really made me think.

In my early teens I came to the view that science was profoundly anti-religious, that it offered a total explanation of all things, and therefore there was simply no room for God at all. Of course, growing up in Northern Ireland didn't really help things very much. The religious conflicts that I saw between Protestant and Roman Catholic created in my mind the impression that religion always causes conflict. And so I gradually came to the view, which was very widespread in the 1960s, that if there were no religion the world would be a better place. I thought that scientifically it made sense and morally it made sense, and therefore I became one of these atheists who advocated the abolition of religion. In fact, when I read Richard Dawkins, I feel quite nostalgic, because that's the way I used to be!

I studied math and science in high school and then went off to Oxford to study chemistry. That was a very significant time for me, because I began to realize two things. One was that the intellectual case for atheism was much weaker than I had imagined. More significantly, I began to realize that the sciences had their limits and that those limits had to be identified and respected. This line of thought started in the school science library, where I accidentally came across the history and philosophy of science section. Up to that stage I'd been a "scientific positivist:" I thought that what science proves is okay and nothing else matters. When I started to study the history and philosophy of science it really brought home to me that actually things weren't that simple. It did not make me believe in God but it certainly began to sow the seeds of doubt about scientific positivism. The philosopher Daniel Dennett talks about what he calls the "universal acid of Darwinism," meaning it's something that erodes everything. For me, however, the history and philosophy of science is an acid that erodes the simplistic "scientific positivist" approach to science that we find in Dawkins and others.

When I left high school I was still an atheist but I was an atheist who was beginning to appreciate that things were not quite as simple as I had thought. I went to Oxford really wanting to keep these questions open because I wasn't quite sure I'd got them right but not actually expecting that anything would happen. I didn't expect to change my mind about being an atheist; that was off limits as far as I was concerned. My impression was that I'd review the evidence and then say, "Right, that's sorted out, let's close this down." But actually it didn't happen like that at all, which was quite disconcerting. Outside of my studies I was thinking, talking to people, reading books, going to lectures, and being bombarded with challenges. I met some very articulate Christians who were able to show that their faith made sense and challenged the shallowness of my thinking. I heard some lectures by the theologian Michael Green, who was a very significant influence because he articulated a vision of Christianity that really made sense. These things forced me to keep my mind open. They sowed more seeds of doubt about atheism and also began to open up possible alternatives. It's one thing to be shown that your worldview actually doesn't make sense but it's quite another to say there is something that's better. I was at a transitional stage. I began to realize that I had a huge amount of re-thinking to do and this eventually led to my becoming a Christian.

I can't put my finger on a definite date or time when I became a Christian. It's not as if suddenly there was a devastating moment of illumination or "road to Domestos[1]" experience as the British comedian Peter Kay would say. What I know is when I went to Oxford in October 1971 I was an atheist but not a happy atheist. When I went home for my Christmas vacation two months later I'd become a Christian. At some point in between I changed my mind.

When I went back to Oxford after that first vacation, I went to see my tutor Jeremy Knowles, who is now Professor of Chemistry at Harvard, and said, "I think I want to change to theology." I realized that faith was intellectually very exciting and I wanted to study the detail. He wrote me a letter laying out why I shouldn't. He said, "Look, you're a good scientist. Don't give it up; see it through to the doctoral level and beyond. If this is really what you want to do then do it but please don't give up the science just yet. You want to really master your discipline first before changing and Oxford's a very good place to master it, so please stay here and do this." So that's what I decided to do. It was quite a big decision but my tutor was right.

When I became a Christian I had to go through a process of rearranging my mental furniture, which was quite a challenging process, and it did take me a while to get my bearings here. My initial instinct was to not think about the relationship between science and faith. One option that some of my friends recommended was keeping them in separate, watertight compartments. I realized very quickly that this was extremely stupid and actually dangerous, because these two things could end up creating tension, where my faith led me one way and my science led me another way, and I could find myself in real difficulty at some point in life. I felt isolated: there weren't many people around who were interested in talking about these things – or at least if there were I didn't know who they were. That was one of the most frustrating things – this awareness that there were so many questions and the feeling that there was nobody really I could talk to about them. There were probably two main issues: one was that I realized I did not have a good enough grasp of my faith and therefore I was engaging with issues in science and religion on the basis of a less than fully developed way of thinking. Secondly I was aware that there were huge areas of science which I didn't really understand. I felt that I was just linking one very small area of science to my faith which I didn't really understand properly, so I did feel very discouraged because I couldn't see quite how I was going to sort this one out. One of the people who helped me was Charles Coulson, Oxford's first Professor of Theoretical Chemistry, who was also a Methodist lay preacher, and based at Wadham College where I was. He always emphasized that faith doesn't reside in gaps in knowledge but that faith is all about a "big picture" way of looking at things. I took that away and thought about it for a long time. He gave me the desire to go away and do some work on this subject myself.

At that time I was helped by the thought that if God made the world, then studying it might actually help me to know more about God or make me appreciate God more. I began to find that the natural sciences actually resonated with my faith. For me one of the enormous attractions of Christianity is its capacity not simply to make sense by itself but to make sense of things. To my way of thinking, it makes an awful lot of sense of the sciences, why they work, what they're saying. There's a quote from C. S. Lewis that I didn't know at that time but I wish I had known it when I was doing research: "I believe in Christianity as I believe the sun has risen, not just because I see it but because by it, I see everything else."[2] I began to realize that when faith is properly understood it has the capacity to illuminate just about every aspect of life, including science. We have nothing to fear from the truth.

A Different Calling

While I was studying science I was reading theology in my spare time. Towards the end of my first year as a PhD student in Oxford, I realized that I couldn't go on like this, I had to make the switch from amateur to professional theology. My agenda was to try to engage properly with the sciences as a theologian. At that time I was offered a PhD scholarship by Merton College but when I read the small print I realized it was *either* for research *or* a second first degree. I asked if I could do both, and although I think the college's governing body thought I was slightly mad, they agreed. So while I was a PhD student carrying out research in biochemistry I spent most of my daylight hours working in the lab and the evenings doing theology.

I then left the world of scientific research and went to Cambridge University to study theology in more detail and prepare for ordination in the Church of England. At quite an early stage as a Christian I had felt that the right thing was to go into Christian ministry. Nowadays I would not be so sure about that. I would now want to say very clearly to anybody who is going into the sciences and sees this as being a long-term vocation that they should see that as being called to a form of Christian ministry. There's a real need for Christians who are very committed to their discipline, and who can bring a Christian perspective to bear on it. We need them to be supported by the church as a whole. Also, when they get to an appropriate point in their careers, they can be mentors to those who are coming after them and are asking some of the questions I asked when I was younger.

I was very clear that if I was going to do theology then I wanted to do parish ministry, because that was very much about putting these ideas to work. The next step for me was to spend three years in parish ministry in Wollaton, a suburb of Nottingham. This turned out to be very good for me because it forced me to appreciate the importance of communication skills. Preaching is an art form and I quickly discovered I wasn't really very good at it. People would very often say to me after a sermon, "You clearly wanted to say something very important but we weren't quite sure what it was!" Those three years in Nottingham were very significant because they gave me the space to develop those skills. Throughout my career since I have kept that real desire to develop a capacity to connect with our culture in terms it can understand. There's simply no point in using traditional Christian language to an audience that doesn't understand it.

My thought had always been to go back into academic life in some shape or form, and so I became a tutor in Wycliffe Hall at Oxford University and stayed there for the best part of twenty-one years. That gave me the opportunity to start doing theology very seriously in a rigorous academic context but also forced me to ask how best to link theology up with real life. Dawkins tries to portray theology as a non-subject, but what I discovered was that theology is about the rigorous study of what Christians believe, looking at the grounds of those beliefs, the nature of those beliefs and the implications of those beliefs. That's a real subject. Theology is becoming more and more important, as the number of people studying it shows. Religious studies is booming in the UK in quite a big way, because it deals with real questions that people really want to explore.

There is no doubt that our culture is still interested in the God-question. My concern is that someone who is academically rigorous is often seen as being isolated from culture at large. There are quite a few of us who believe that academic excellence offers a robust foundation which enables us to engage cultural questions with confidence; not just on the basis of, "I feel this" but actually, "I think I can make a case for thinking like this." When I read the latest novels, and watch movies and TV, I'm always looking for "signals of transcendence." In other words, the questions people are asking and the language they're using is often saying, "We're looking for meaning, there's something significant in life and I haven't found it yet." Then I can begin to phrase what I think in language that is going to be accessible to these people: something Christians would call "apologetics." Eventually I and one or two others began to realize that there was a real need for apologetics to be done intentionally with a very rigorous academic base so that we could actually engage people like Richard Dawkins. That moved us to set up the Oxford Centre for Christian Apologetics at Wycliffe Hall, where people can come and build up their academic foundations and communication skills in order to be able to do this.

Faith and Science

People often don't know enough about the sciences to be able to helpfully make a connection between faith and the world of science. In looking at nature it's not the gaps that are really interesting. The amazing fact is that we can make sense of it at all: it's a big picture view. And so for me, the key lies in that quote from C. S. Lewis, "I believe in Christianity as I believe the sun has risen, not just because I see it but because by it, I see everything else."[3] What Lewis is saying is that if you believe in God properly and rightly, then it makes sense of the whole thing – of everything that we observe.

There are specific interactions with faith in different areas of science; I think it's fairly widely agreed that in astronomy and physics these questions are so obvious that actually people do talk about them, in biology it's much more sensitive, and in psychology there's some real hostility. That's one of the reasons why you need a specialist in each discipline who can help others to cope with the issues there. What I would like to see happen is for people to encourage those who are in significant academic or educational positions to work out how to handle these questions so that they can mentor those who are emerging in those fields. We can't presume that there will be people able to handle these issues in the future, and we've got to try to make that happen.

One of my concerns now is that when they come up against these issues a lot of Christians say, "I don't want to think about this." Very often after several years they do start to think, and because they have no one to help them, they implode. You've got to make sure that you can connect your faith with your professional activity, because in the end one is going to give unless you make sure that they talk to each other. My advice to a student studying any discipline, but especially science, is to find older people in the same field who are Christians and ask them how they connect their faith and their work. These are real questions and very often they're questions that, once answered, lead to a deeper form of faith. I think one of the great things about Christians in Science[4] is making available networks to take this forward, which I did not have access to at a very formative period in my life.

Does Science Point to God?

When science comes into the discussion with people who aren't Christians I very often find people use it as a screen to avoid engaging with any question about God. What I find very disturbing is the way in which a number of scientists, often with very aggressive atheist agendas, smuggle in a whole series of secular philosophical ideas as if they were authentic science. Once you begin to challenge that, the atheist case begins to collapse. However, this kind of approach is very widespread and influential in today's culture. I think there are genuine questions being asked as well. For example, can one prove the existence of God scientifically? Richard Dawkins says that if you can't explain something scientifically, there is no reason to believe it. My response is that there are many things that are enormously important that cannot be shown scientifically.

Peter Medawar, who won the Nobel Prize for medicine, argued that science is very good at explaining the material world but when it comes to questions of value or meaning it's not going to get you very far at all. For most of us, the question of what is right and wrong is very important, and if science can't tell us that, then we go and find out some other way. We can also ask exactly what science has proved, emphasizing that clarification of a mechanism is not the same as establishing the meaning. We know how human life comes about but that doesn't tell us what its meaning is. If I were to point at something, a scientist could give a very good description of how various nerve cells were activated and my muscles moved my hand, but all they would have done is to describe the mechanism by which I pointed. They couldn't tell you what the meaning of the action was. "How" and "why" are very different questions! Science has its limits, and there is a whole range of questions, including the God-question, which lie beyond the range of science. That doesn't mean they're stupid questions, it just means science can't really answer them.

Christians are often so preoccupied with dealing with objections coming from science that they don't give enough thought to how they might use science to make a positive argument for the existence of God. So the question then would be how *could* one use science to do this? For example, science is obliged to presuppose many fundamental assumptions that it cannot prove. Yet many of these are integral to belief in God. For example, the ordering of the world that we see is very significant, as John Polkinghorne (the particle physicist-turned Anglican priest) has pointed out. Why can we make sense of the world? Maybe you can say that our ability to do this just happened by accident so we're just very lucky. Or there might be some deeper explanation here, and most scientists I know are very happy to consider that.

In my own work I have been looking at ways of using nature to point to God, and trying to show how the Christian worldview gives a much more satisfactory explanation of what we know about the world than any of its alternatives. For example, life is only possible because the fundamental properties of the elements (the basic chemical building blocks of everything we see) are the way they are. If the chemistry of the element carbon were very slightly different, there would be no DNA (the information-carrying molecule that programs all the activity in every living cell). There would be no mechanism for the transmission of genetic information, so actually evolution wouldn't happen at all. There is a whole range of things like this. You could simply say that this world seems to have been set up for us to be here.

I chose Joseph Addison's famous "Ode" because it is a classic example of a way of seeing and appreciating nature. Addison was an enthusiastic observer of nature and valued its beauty in its own right. However, he was also very clear that the natural world had an additional layer of meaning. It acted as a signpost to its Creator. Nature proclaims, displays and publishes the identity and glory of its Creator. Addison's poem can be seen as an extended commentary on Psalm 19:1: "The heavens declare the glory of the Lord." As an experienced journalist, Addison was deft with words, allowing his poem to convey both the intellectual and spiritual aspects of reflecting on the natural world. I still read the poem with a sense of delight, enjoying both Addison's poetry and the ideas it expresses.

> The spacious firmament on high,
> With all the blue ethereal sky,
> And spangled heavens, a shining frame
> Their great Original proclaim.
> Th'unwearied sun, from day to day,
> Does his Creator's powers display,
> And publishes to every land
> The work of an Almighty Hand.
> Soon as the evening shades prevail
> The moon takes up the wondrous tale,
> And nightly to the listening earth
> Repeats the story of her birth;
> While all the stars that round her burn
> And all the planets in their turn,
> Confirm the tidings as they roll,
> And spread the truth from pole to pole.
> What though in solemn silence all
> Move round the dark terrestrial ball?
> What though no real voice nor sound
> Amid the radiant orbs be found?
> In reason's ear they all rejoice,
> And utter forth a glorious voice,
> Forever singing as they shine,
> "The hand that made us is divine."

Joseph Addison, "Ode to the glory of God," first published in *The Spectator*, no. 465, 1712. Inspired by Psalm 19.

Epilogue

These interviews were collected over almost a year, from September 2007 to July 2008. During this time there was plenty happening in the media to keep people talking about science and religion, and we didn't have any trouble finding Christians working in science who were willing to be interviewed. Some I met for the first time at the interview, others I had heard give lectures and a few were friends from my time with CiS. It was great to be able to talk to most of these people in their "home" environment. I caught up with Bill Newsome in his office at the end of a long day in the lab, finishing off work before heading to a conference the next morning. I met Jennifer Wiseman for dinner in Washington, in a restaurant that happened to have a view of Capitol Hill, where she used to work. I interviewed Alister McGrath in his Oxford study, a tower room above the chapel at Harris Manchester College, surrounded by piles of books and papers. It was a real privilege to be able to hear about people's lives and some of the events and influences that shaped them as people.

There are several moments from these interviews that have stayed with me and some common threads through all of them that have caught me by surprise. I often remember Rosalind Picard telling me how she learned not to fear other people, and just followed the path her research was taking – even though she knew that her work might not be accepted by some people at first. For many of these scientists there was often a point in their career where they changed track unexpectedly or focused their research in a new way. Ard Louis decided to follow riskier but more meaningful long-term projects rather than research that would result in more instant publication and recognition. Jennifer Wiseman turned down a university faculty position to spend time working with the US Senate before heading back into research, and John Polkinghorne gave up a successful career in science to be ordained. It's obvious from what they have said that their faith has led them to, or enabled them to make, these risky decisions. The world is a richer place because their faith provoked them to think about things in a different way.

Something that also struck me is that many of the scientists I spoke to have thought hard about whether what they are doing is worthwhile. Alasdair Coles tried working in a developing country, Deborah Haarsma was relieved when she realized that science could be a worthwhile "calling" and Ard Louis at first saw science as a means to an end, rather than being of use in itself. Some of them thought about other career plans but they came to the same conclusion as Bill Newsome: that they were already using their talents in the best way they knew how (for the time being), in science.

In terms of their faith, you can see very clearly how logical minds find evidence for the claims of Christianity, weigh them up, and test faith with experience. C. S. Lewis's famous book *Mere Christianity* was key for several of them. In it he lays out the evidence for Christianity, and comes to the logical conclusion that a step of faith is not an irrational thing at all. But this isn't the only factor. Other things have been important: personal experience, relationships, historical evidence and, most importantly, God's revelation of himself through Jesus.

The unique thing about all of these people as Christians is that when they are amazed at the beauty and intricacy of nature they, as Deborah Haarsma has said, "see a connection back to who the Creator is."

For some, this may have been a rare look at how Christians who are also working scientists think about their faith and science together, and how this impacts on their work. You may not agree with all of what has been said but hopefully they have proved that it is possible for a successful scientist to be a person of faith. They may also have showed some new ways of looking at the world. If you're left thinking, "But what about ...?" then as well as the *Test of Faith* materials that accompany this book, several of the people featured in these pages have written books that take their ideas further.

The question is, "Where to from here?" It's fascinating that science has such authority in our culture. People are looking for evidence for or against God, for meaning, for their roots or for a place in the world. It's important to keep on showing how science makes sense in the light of Christian faith. We hope that when scientists like these tell their stories it will become increasingly difficult for people to claim that science and religion are at war. It's important for more and more Christians working in science to be able to talk about their faith and how that relates to their work, and to realize what an impact those stories can have in breaking down old myths and barriers. Bill Newsome's story about giving his first science-faith lecture illustrates this process so well.

As new discoveries are made, there are always new questions that people of all faiths and none should be asking: What are the best ways to use the latest technologies? How do we function as human beings? What is so special about us? How should we use the resources available to us? The Christians who are involved in science can help to make sure that these issues are discussed in ways that are relevant to the millions of others who share their faith but not their scientific knowledge. Churches also have a big role to play in encouraging that conversation to happen.

For those readers who are Christians already working in science, hopefully hearing these stories has inspired you to keep going in your research or teaching; to think about things in a different way; to delve more deeply into how your faith relates to the work you're doing; or to tell your own story. The two organizations that have been mentioned, the American Scientific Affiliation and Christians in Science, are a great place to start.

Alister McGrath mentioned the need for Christians working in science to mentor students and help them to integrate their faith and their studies. This is possibly the most important factor in the whole equation. The ones who will be doing the most important thinking in this area for the rest of this century are in school, college or university now. If the people around them can help them to think more clearly about science-faith issues and start to ask new questions, then that's a very worthwhile investment.

I will finish with a prayer from a well-known astronomer. I have chosen it because I think it really sums up what the scientists in this book have been saying. Johannes Kepler was the first person to accurately describe the orbits of the planets in our solar system, and "Kepler's laws" are still used by astronomers today. Kepler was also a Christian, and he dedicated one of his books with this prayer:[1]

> I give you thanks, O LORD and Creator, that you have gladdened me by your creation, when I was enraptured by the work of your hands. Behold, I have here completed a work of my calling, with as much intellectual strength as you have granted me. I have declared the praise of your works to the men who will read the evidences of it, so far as my finite spirit could comprehend them in their infinity ... Have I been seduced into presumption by the admirable beauty of your works or have I sought my own glory among men, in the construction of a work designed for your honor? O then graciously and mercifully forgive me; and finally grant me this favor, that this work may never be injurious but may lead to your glory and the good of souls.

Endnotes

Forword

[1] dense.

Introduction

[1] A UK-based organization for Christians working in or studying science, and others who are interested in the interaction between the two. www.cis.org.uk

[2] Where about 84% of people declare themselves as Christian and 47% attend church, compared to 72% and 15% in the UK. (Statistics from *The State of the Church: 2006;* Barna Group; *Churchgoing in the UK,* Tearfund 2007; UK government census 2001.)

[3] Edward J. Larson and Larry Witham, "Scientists are still keeping the faith," *Nature* 386:435–436 (3 April 1997).

[4] The 39% of scientists who believe in a personal God who answers prayer roughly reflects the 45% of US citizens who have made a personal commitment to Christianity (Barna Group).

[5] Larson and Witham, *Nature* 394:313, 1998.

[6] E. H. Ecklund, "Religion and Spirituality and University Scientists," Social Science Research Council Essay Forum, Feb. 05, 2007.

[7] P. L. Yeyati, R. M. Bancewicz, J. Maule and V. van Heyningen. Hsp90 "Selectively Modulates Phenotype in Vertebrate Development." PLoS Genetics 2007, Mar. 30; 3(3):e43.

[8] John Stott, *The Contemporary Christian: Applying God's Word to Today's World* (Chicago: InterVarsity Press, 1995), emphasis mine.

[9] People who use science to defend an aggressive atheist agenda, including Richard Dawkins, Christopher Hitchins, Daniel Dennett, Peter Atkins, Lewis Wolpert and Sam Harris.

Chapter 1

[1] The mathematics of atoms and molecules, and the particles that make up atoms (such as electrons, protons, quarks and gluons).

[2] Deism is the belief that a god started the universe off but has not been involved with it since.

Chapter 3

[1] Dr. Wiseman is representing her own personal views and not those of NASA.

[2] A university that has produced 72 Nobel Prize winners.

[3] The fundamental physical principles governing atoms and molecules, and the particles that make up atoms (such as electrons, protons, quarks and gluons).

Chapter 4

[1] Government-subsidized housing.

[2] A UK-based movement, now renamed Urban Saints.

[3] She was chair of the Human Fertilisation and Embryology Authority (HFEA) from 2002 to 2008. The HFEA is the regulatory group for all work involving embryos, IVF and donor insemination in the UK. http://www.hfea.gov.uk

Chapter 5

[1] In John 9.

[2] A student Christian movement. http://www.intervarsity.org/

[3] An organization that encourages students to think about how faith relates to academic issues. http://www.veritas.org/

Chapter 6

[1] Pastor of Park Street Church in Boston.

Chapter 7

[1] A UK-based organization for Christians working in or studying science and others who are interested in the interaction between the two. www.cis.org.uk

[2] The Faraday Institute for Science and Religion is an academic research enterprise based at St. Edmund's College, Cambridge, England. www.faraday-institute.org

Chapter 8

1. This is the title of John Polkinghorne's autobiography, published in 2007 by SPCK.
2. Luke 19:1–10. A man who collected taxes for the Romans climbed a tree so that he could see Jesus over the crowd. Jesus saw him and invited himself to his house that evening. Zacchaeus was so impressed by Jesus' teaching that he paid back all the people he had cheated, with interest.
3. The study of the parts that make up atoms: quarks, gluons and protons, etc.
4. An ordained role that includes many of the duties of a vicar or curate (trainee vicar).
5. Thirty-nine to date. Details on http://www.polkinghorne.net/
6. A whole range of factors are just perfect for life to develop in the universe: gravity, electromagnetism, the forces that hold the atom together, and so on. If any of them were the tiniest bit different, we wouldn't exist.
7. Where our universe, like Goldilocks's favorite chair, happens to be fit for life to develop.
8. The idea is that if every action of ours were simply the result of a physical chain-reaction in response to an event, then we would not have freewill.
9. A single equation uniting the different physical forces: nuclear, electromagnetic and gravitational, that would help scientists to understand more about how the universe is the way it is.

Chapter 9

1. See www.random-art.org

Chapter 10

1. Domestos is a popular brand of household bleach in the UK.
2. *Is Theology Poetry?* (1945).
3. *Ibid.*
4. A UK-based organization for Christians working in or studying science and others who are interested in the interaction between the two. www.cis.org.uk

Epilogue

[1] From the end of the final chapter of "The Harmony of the Worlds,"
Volume 5 (the volume that describes the third law of planetary
motion). This version is adapted, with permission, from the one
quoted in D. Alexander and R. S. White, *Beyond Belief: Science, Faith
and Ethical Challenges* (Oxford: Lion, 2004), pp 56–7.

Test of FAITH

SCIENCE AND CHRISTIANITY UNPACKED

Is God a delusion? Has science removed the need for faith? How do we relate to scientific discoveries?

The creative and visually stunning three-part documentary **Test of FAITH** uses interviews with top scientists and theologians to introduce the idea that science and faith do not have to be in conflict.

The documentary also introduces some hot topics:

- WHY ARE WE HERE?
- HOW WAS THE UNIVERSE MADE, AND WHY?
- WHY SHOULD WE CARE FOR OUR ENVIRONMENT?
- WHAT GOES ON INSIDE OUR BRAINS?
- WHAT DOES BEING MADE IN THE IMAGE OF GOD MEAN, AND HOW DOES THAT AFFECT MEDICAL ISSUES LIKE GENETIC TESTING?

The **STUDY GUIDE** then allows small groups to unpack these issues, and discuss them at a level and pace that suits the group. It's flexible so that you can choose the topics you want to cover.

The **LEADERS GUIDE** provides all the content of the study guide plus suggested responses to questions, critical background information, and opportunities for taking these issues further.

www.testoffaith.com

Introductory resources from **The Faraday Institute for Science and Religion**

A range of resources designed to help you explore **science-faith issues**

- **VIDEOS**
- **ARTICLES**
- **STORIES**
- **REVIEWS**
- **MATERIALS FOR CHURCH SERVICES**
- **EXTRA RESOURCES TO ACCOMPANY THE TEST OF FAITH COURSE**